MAPPING MEN AND EMPIRE

Adventure stories, produced and consumed in vast quantities in eighteenth-, nineteenth- and twentieth-century Europe, narrate encounters between Europeans and the non-European world. They map both European and non-European people and places. In the exotic, uncomplicated and malleable settings of stories like *Robinson Crusoe*, they make it possible to imagine, and to naturalise and normalise, identities that might seem implausible closer to home. They make it possible to map new forms of masculinity, as writers such as Robert Ballantyne sought to do. At the same time, adventure stories chart colonies and empires, projecting European geographical fantasies onto non-European, real geographies, including the Americas, Africa and Australasia.

But beneath the map-like realism of adventure stories, there is an undercurrent of ambivalence. Adventure's geography is more fragile and also more fluid than it first appears. While adventure stories map, they also unmap geographies and identities, destabilising and sometimes recasting them. The ambivalent geography and politics of adventure are illustrated in late-Victorian and Edwardian girls' stories, in which boundaries between masculinity and femininity are blurred, and in contemporaneous stories by Jules Verne, which can be read as anarchist adventures.

The ambivalent geography of adventure has potential to be a site of resistance. Writers and readers have appropriated adventure, and found a metaphorical space in which to challenge hegemonic constructions of identity and colonial geography. Presenting what Robert Louis Stevenson (author of *Treasure Island*) called a 'kaleidoscopic dance of images', the geography of adventure is alluring, fluid, liminal and open, a space in which people and places are continuously mapped, unmapped and remapped.

Richard Phillips is lecturer in Geography, University of Wales, Aberystwyth.

MAPPING MEN AND EMPIRE

A geography of adventure

Richard Phillips

Routledge
Taylor & Francis Group
New York   London

First published 1997 by
by Routledge
2 Park Square, Milton Park, Abingdon, Oxon OX14 4RN

Simultaneously published in the USA and Canada
by Routledge
711 Third Avenue, New York, NY 10017

Routledge is an imprint of the Taylor & Francis Group, an informa business

© 1997 Richard Phillips
Typeset in Garamond by Keystroke, Jacaranda Lodge, Wolverhampton

British Library Cataloguing in Publication Data
A catalogue record for this book is available from the British Library

Library of Congress Cataloguing in Publication Data
Phillips, Richard.
Mapping men & empire: a geography of adventure / Richard Phillips.
p.   cm.
Includes bibliographical references and index.
1. Adventure stories, English—History and criticism.
2. Adventure stories, Australian—History and criticism.
3. Adventure stories, French—History and criticism. 4. Defoe, Daniel,
1661–1731. Robinson Crusoe. 5. Intercultural communication in literature.
6. Masculinity (Psychology) in literature. 7. Difference (Psychology) in literature.
8. Defoe, Daniel, 1661–1731 Influence. 9. Imperialism in literature. 10. Geography
in literature. 11. Colonies in literature. 12. Boys–Books and reading. 13. Men–
Books and reading. 14. Travel in literature. I. Title.
PR830.A38P48 1996
823′.08709–dc20   96–10899

ISBN 0–415–13771–3
0–415–13772–1 (pbk)

CONTENTS

PLATES

ACKNOWLEDGEMENTS

Behind this book – about everyday geographical imagination and the places it can take us – are some wanderings of my own, which began in 1989. In August that year I arrived at Vancouver International Airport. Vancouver was to be my base for the next five years.

Cole Harris appeared at the airport, and soon drove me to a rocky point, which seemed to prove *his* point: we were perched on the edge of the wilderness. Wilderness, perhaps not, but this was at least a good point of departure. . . . Cole was to be critic, editor, friend and – as I learned to my peril – leader of field trips and singer of folk songs. Others at the University of British Columbia influenced and supported me in other ways, also in the best Vancouver traditions. Over Starbucks coffee, Bill New showed me something of Canadian and Australian literature, and gave me some important criticism and advice. Over (deceptively strong) Canadian beer, Derek Gregory introduced me to geography more vital and more open than I had known before. And between gin and tonics, Sheila Egoff told me about Canadian children's literature.

It was also at UBC that I met Alison Blunt, David Demeritt, Joy Dixon, Graeme Wynn, Mark Duffet, Katy Pickles, Martin and Juliet Rowson-Evans, Miguel Lopez, Stacy Warren and Yas Quereshi, many of who became friends, and all of whom influenced and contributed to my work. I also benefited from conversations with friends outside the world of the University. In particular, I am thinking of Adrienne Webb, Chris Woodland, Howard Kwan, Joan Calderhead, Michael and Mona Smayra, Rachana Raizada and Sherman Levine, all in Vancouver. I spent quite a lot of time on the road, both in Canada and overseas, including Australia – where I met Dennis Jeans, Eli Franco, Kay Anderson, Michael Cathcart, Paul Carter, Ross Gibson and others, all of whom left their mark on what I have written.

I came to Wales in 1994, bringing a draft of the book with me, which I have since revised. I am grateful to those who have helped, particularly Vaughan Cummins, also Felix Driver and – once again – Alison Blunt, Cole Harris, Derek Gregory and Bill New. I am grateful to John Lewin and Robert Dodgshon at the University of Wales, and Tristan Palmer and Sarah Lloyd at Routledge, for their confidence in me and in this book. Also to other editors at Routledge, including Matthew Smith, Caroline Cautley and Lindsey Brake, who have also worked hard to see the MS into print.

I would like to thank the library and technical staff who have assisted with the research, particularly those at Bethnal Green Museum of Childhood (Tessa Chester), the British Library and Newspaper Library (including Jill Allbrooke), the Mitchell Library at Sydney, the National Library of Wales, the Public Archives of Canada (Ottawa) and Provincial Archives of Saskatchewan (Regina and Saskatoon), the University of British Columbia Special Collections, Vancouver Public Library (Terry Clark), and the University of Wales, Aberystwyth (notably David Griffiths).

Finally, I am privileged to have been supported by the Government of Canada, in the form of a Commonwealth Scholarship, by occasional research funds from Cole Harris, and by awards from the University of Wales, Aberystwyth.

PLATES

Permission is gratefully acknowledged to reproduce the following plates:

Plates 1.4 (12613.bb.6), 2.1, 4.1 (012808.h.38), 4.2 (2374.g.2), 4.3 (012808. h.38), 4.4 (012808.h.38), 5.4 (012804.i.44), 6.1 (838.a.1), 6.2 (12511.ff.31) and 6.3 (12511.ff.31) by permission of The British Library; Plate 1.5 by permission of Cambridge University Press; Plates 2.3 and 3.2 by permission of the National Library of Wales; Plate 2.4 courtesy the Boston Museum; Plate 7.1 by permission of Orion Publishing Group; Plate 7.2 courtesy M. Tournier (1972) *Friday and Robinson*; Life on Speranza Island, New York: Knopf; Plate 8.1 copyright House of Viz/John Brown Publishing.

Every attempt has been made to obtain permission to reproduce copyright material. If any proper acknowledgement has not been made, we would invite copyright holders to inform us of the oversight.

1

INTRODUCTION
Adventures in the New World

'The literature of geography', American geographer Carl Sauer once observed, 'begins with parts of the earliest sagas and myths, vivid as they are with the sense of place and of man's contest with nature' (Sauer 1925: 21). This literature extends, in modern form, to the narratives of explorers, surveyors, geographers and other storytellers, who describe journeys 'into the unknown' – adventures.[1] Since the eighteenth century, the eclectic literature of adventure has been consumed by mass reading audiences in Europe; it has fed geographical imaginations. Adventures have charted cultural space with profound and far-reaching implications, both for those who imagine the 'unknown', in which stories are set, and also for those who live there, in geography unknown to others but known to themselves.

POINTS OF DEPARTURE: SKETCHY MAPS

Now when I was a little chap I had a passion for maps. I would look for hours at South America, or Africa, or Australia, and lose myself in all the glories of exploration. At that time there were many blank spaces on the earth, and when I saw one that looked particularly inviting on a map (but they all look like that) I would put my finger on it and say, When I grow up I will go there.

(Conrad 1899: 197)

Marlow, Joseph Conrad's adventurer in *Heart of Darkness* (1899),[2] recalls his boyhood fantasies of adventure, which were freely accommodated by the blank spaces he found on maps. Generations of adventure writers, heroes and readers have been inspired by sketchy maps, both real and imaginary, which seem to invite their geographical fantasies. Robert Louis Stevenson was inspired to write the adventure classic *Treasure Island* (1883) when he was gazing at the map of an imaginary island he had drawn for his young stepson (Plate 1.1).[3] The map, which appears in his story, depicts an island 'Offe Caraccas', and includes some cryptic directions to buried treasure. It shows 'latitude and longitude, soundings, names of hills, and bays and inlets, and every particular that would be needed

1

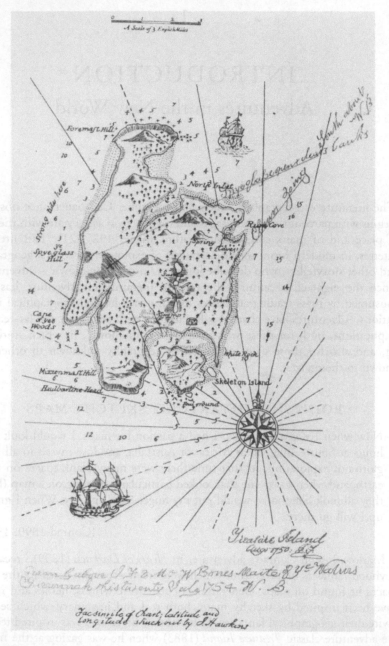

Plate 1.1 Map illustration showing Robert Louis Stevenson's Treasure Island (1883). Originally printed in colour, this was the only illustration to appear in the first book-length edition of *Treasure Island*. An alluring and ambiguous geographical suggestion.

to bring a ship to safe anchorage there' (Stevenson 1883: 51). But the map is sketchy, with few formal cartographic symbols and minimal detail; it is without scientific pretensions. It is, rather, an alluring and ambiguous geographical suggestion. 'Brief' though it is, and 'incomprehensible' to the young hero, Jim Hawkins, the document fills the adventurers 'with delight' (Stevenson 1883: 51–52). Jules Verne, too, is said to have dreamed up adventure stories while gazing at maps and charts (Aitchison 1909: 557). Many adventures have begun in this way, as outline maps (like Jim's), or *terra incognita* on larger maps (like Verne's), chart spaces in which anything seems possible and adventure seems inevitable.[4] In these malleable spaces, writers and readers of adventure stories dream of the world(s) they might find, the adventures they might have, the kinds of men and women they might become.

The dream of *terra incognita* appears, for a moment, both universal and innocent. Universal, it is a dream we have all dreamed. As a President of the Association of American Geographers (AAG) once put it,

> Terra incognita: these words stir the imagination. Through the ages men have been drawn to unknown regions by Siren voices, echoes of which ring in our ears today when on modern maps we see spaces labelled 'unexplored,' rivers shown by broken lines, islands marked 'existence doubtful'.
>
> (Wright 1947: 1)

Another AAG President has diagnosed the fascination with real and imaginary maps as a common, perhaps universal condition: cartophilia (Lewis 1985: 465). The dream of *terra incognita* also seems innocent – both harmless and private. Conrad's image of a *child* gazing at a map and dreaming of adventure invokes late-Victorian associations between adventure and children's literature, and between children and innocence.[5] Marlow's childhood dreams seemed innocent, but his innocently white spaces of 'delightful mystery' were later coloured by experience; they became the 'dark spaces' of a brutal adult world. His seemingly personal dreams of adventure connected with social worlds and social politics, in particular the politics of European imperialism in Africa and imperial masculinity in Europe. Despite parallels with universal or mythical fantasies and narratives (Campbell 1949; Torgovnick 1990), Marlow's dream was specifically the dream of a white boy, growing up in an age of empire.

European empires and European masculinities were imagined in geographies of adventure. *The North-West Passage*, painted by British Pre-Raphaelite artist John Everett Millais in 1874, illustrates the association between British imperialism, British imperial masculinity and British dreams of adventure in unknown geography (Plate 1.2). Despite his title, Millais made no attempt to paint a 'real' North-West Passage. Instead, he pointed to a region of the imagination, contemplated by an old explorer, whose thoughts and gaze rest on the horizon, rather than in the room where he sits with a young woman. The old man dreams of distant lands and uncharted seas. A telescope and an open

3

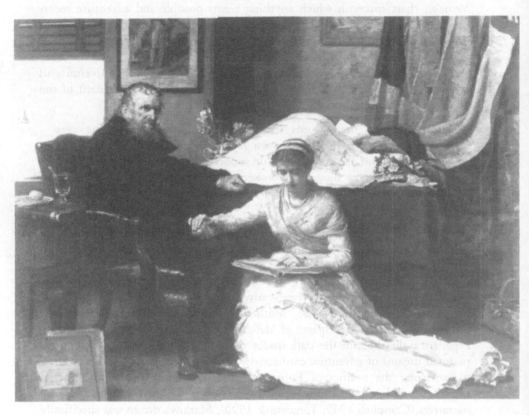

Plate 1.2 The North-West Passage by John Everett Millais, 1874. Imaginative geography, accommodating constructions of masculinity and imperialism.

window draw him into the distance, where a sailing ship disappears out of view, suggesting movement and escape. The distance is hazy, uncharted but alluring, not only in the window, but also in the sketchy map lying on the old man's desk, and in the partially veiled painting of icebergs, in which another sailing ship fades into the distance, luring the imagination to geographies of adventure. As the Dutch map-maker Joan Blaeu wrote, 'maps enable us to contemplate at home and right before our eyes things that are farthest away'.[6] The imaginative geography of Millais' *North-West Passage* is explicitly tied to constructions of masculinity and imperialism. The geographical fantasy belongs to the man. While he is drawn to distant lands and seas, the woman is absorbed in her book, content in the room where she sits, and oblivious to the old man's restlessness. Nothing could be further from her mind than the North-West Passage. The old man's geographical fantasy, not just a dream, is associated with real acts of imperialism, partly through the picture of Nelson above the explorer's head and the British flags draped across the wall. In case any viewer should miss the point, the canvas was exhibited under the quotation, 'It might be done, and England should do it.' The associations between adventure, masculinity and imperialism in *The North-West Passage*, in the view of art critic Joseph Kestner (1995: 28), belong in a 'continuum of heroizing masculinity'.

But while geographical imaginations and adventure narratives often appear committed to 'continuous reinscription' of dominant ideologies of masculinity and empire (Kestner 1995: 29), the geography of adventure is neither deterministic nor static. The open window and sketchy map in Millais' painting are points of departure, which may lead to reinscription of dominant ideology, but may lead somewhere entirely different. The prospect of unknown geography, with open windows and sketchy maps, introduces a dynamic to the system. So, while writers and readers of adventure stories dream of the world(s) they might find, and the kinds of men and women they might become, they sometimes reproduce but sometimes transgress dominant ideologies.

SKETCHY BUT REALISTIC:
MAPS IN MODERN ADVENTURE

Arthur Ransome, author of adventure stories such as *Swallows and Amazons* (1930), sketched out a genealogy of modern[7] British[8] adventure in a reading list intended for children who had enjoyed his own books (Hardyment 1984: 220). The list begins with *Robinson Crusoe*, which Ransome says is 'a very important book for those of you who want to know what to do on a desert island. It is also good about shipwrecks and voyages.' Also included in Ransome's list is Stevenson's *Treasure Island*, R.M. Ballantyne's *The Coral Island* (1858a) and *Midshipman Easy* (1834), by Captain Marryat. *The Coral Island* and *Midshipman Easy*, and to a lesser extent *Treasure Island*, are all Robinsonades, stories modelled on *Robinson Crusoe* (Green 1990). The term 'Robinsonade', from the German 'die Robinsonade', refers to the predominantly European genre of realistic survival and adventure stories, most (but not all) of which were published after *Robinson Crusoe*.[9] Also included in Ransome's list were Robinsonades that readers would find in the adult rather than the children's rooms of their library, including 'the works of Joseph Conrad' and Herman Melville's novels, *Typee* (1846) and *Moby Dick* (1851).[10] Another adventure classic on the list, and one that would be found on the non-fiction rather than the fiction shelves, was Richard Hakluyt's *Voyages* (1598–1600).[11]

Ransome's history of adventure roughly spans the European age of exploration, when Europeans were exploring and mapping what they called the 'New World'. European geographers sketched the general outlines of Africa, the Americas and the 'Southern continent' (which originally included Australia and Antarctica). Between 1400 and 1550, most of the world's coastlines were charted and added to European world maps. The invention of movable type, at around the same time, made it possible for maps – new and old – to be reproduced, hence to make stronger impressions on the popular geographical imagination. Australia and North America presented the most striking blanks on maps of the world, such as Mercator's famous and much-copied 1587 map (Plate 1.3). The Mercator map, based on 'the first modern geographical atlas', as Gerard Mercator (cartographer) and Abraham Ortelius (publisher) billed their *Theatrum Orbis Terrarum*, effectively became the definitive modern European world map (Brown 1977: 162). It established an intellectual framework in which contemporary geographical knowledge was logged, and ignorance made visible. Throughout the course of the sixteenth, seventeenth, eighteenth and nineteenth centuries, European geographers filled the hazy outlines on their sea charts and maps with names and symbols. Towards the end of the nineteenth century, they scrambled to map the remaining blanks, which were concentrated in the African and Australian interiors and in the Canadian Arctic. The Royal Geographical Society (RGS), for example, sponsored and rewarded expeditions to these regions, and published explorers' narratives and maps. By the turn of the century, however, it seemed to many that the world was mapped. It is not, of course, necessary to subscribe to this progressivist, linear view of geographical knowledge today; while they were generally more crowded, successive maps were not necessarily any more truthful (Livingstone 1992). But to many Europeans and Americans, from Jules Verne to Frederick Jackson Turner, the turn of the century seemed to herald the conclusion and the end of geography; there was nowhere left to go.

Ransome's history of adventure also spans the European age of empire building. Imperialism went hand-in-hand with mapping, by which Europeans imaginatively and materially possessed much of the rest of the world, including the 'New World' (Livingstone 1992). Cartographers and other map-makers, including adventure story writers, charted areas of geographical knowledge and *terra incognita*, and through their maps they possessed real geography. In cartographic and literary maps, Europeans charted the world then colonised it (Said 1993). The late nineteenth-century scramble to map was also a scramble

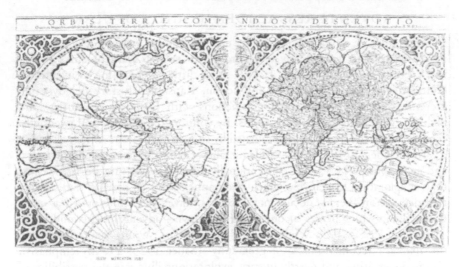

Plate 1.3 Orbis Terrae Compendiosa Descriptio by Rumold Mercator, 1587
(source: Nordenskiold 1973: plate XLVII). Based on a map produced by Gerard
Mercator in 1569, the 1587 map was probably issued as a separate sheet map before
appearing in the Mercator Atlas in 1595 (Moreland and Bannister 1986). Australia
and North America present the biggest blank spaces, partly because the projection
exaggerated the size of extreme northern and southern regions.

to colonise and consolidate imperial power. European imperialism and map-
making reached a simultaneous climax at the end of the nineteenth century.

'By this time,' Conrad's Marlow continued, at the end of the nineteenth
century, '[Africa] was not a blank space any more. It had got filled since my boyhood with rivers and lakes and names. It had ceased to be a blank space of
delightful mystery – a white patch for a boy to dream gloriously over' (Conrad
1899: 197; see Driver 1994). As *terra incognita* disappeared from European
maps, writers of adventure stories retreated from realistic to fantastic, purely
imaginary spaces. Jules Verne wrote the first popular science fiction novels,
removing adventure to settings under the sea and at the centre of the earth, for
example, in *Twenty Thousand Leagues Under the Sea* (1870) and *A Journey to
the Centre of the Earth* (1864). Arthur Ransome set *Swallows and Amazons* in the
British Lake District, not as it appeared on maps, but as it is reinvented by
children, as a space in which to encounter Amazonian Indians. The adventures
of Verne and Ransome, set in essentially unrealistic spaces, signalled the beginning of the end of the realistic adventure story, although realistic adventures
continued to be written and read until well after the end of the Second World
War.

ADVENTURES, ACADEMIC AND POPULAR

The literature of adventure is eclectic, encompassing a variety of adventurers in a variety of narratives. Many adventure stories, such as Daniel Defoe's 'fiction that reads astonishingly like fact',[12] show little respect for conventional boundaries between fiction and non-fiction. Ransome's history of adventure literature mixes relatively sober non-fictions with lively action stories, and includes both adult and juvenile works. By volume, nineteenth- and twentieth-century adventure literature is dominated by so-called children's stories, although it also includes more formal – academic and scientific – texts. Many historical and geographical works, not conventionally labelled adventure stories, are just that. Exploration narratives, which geographers and historians traditionally regarded as non-literary, documentary sources, have been reinterpreted as quest narratives, in which heroes encounter the unknown – adventures (Atwood 1972; MacLulich 1977, 1979; White 1973). 'Explorations,' wrote Sauer, 'have been the *dramatic* reconnaissances of geography' (Sauer 1925: 21, my emphasis). Dramatic, exciting and overtly literary, explorations illustrate the possibility of reading geography as adventure, geographical narratives as adventure narratives. The works of modern geographers such as Sauer can also be read as adventure stories. *The Early Spanish Main* (1966), Sauer's account of Spanish colonisation in the Caribbean, is a case in point.

The Early Spanish Main is a dramatic and engaging retelling of an old adventure story, while it is also a scholarly and critical geographical tome. Like many of Sauer's works, it is a narrative of European encounters in the non-European world, with unknown space and unknown peoples.[13] Like other adventure stories, it is the story of a man – Christopher Columbus – who leaves Europe and ventures into the unknown. Solitary, romantic, brave, impractical, anti-social, masculine, white and European, he is an almost archetypal adventure hero (see Campbell 1949; Zweig 1974). He abandons himself to chance, sailing into uncharted seas in a quest for gold and power. The central image of unknown space in this adventure story is, perhaps inevitably, the sketchy map of an island (Plate 1.4). Sauer tells how no other map has ever been attributed to Columbus. He tells how Columbus failed or refused to map the Caribbean in any detail in his letters – which left the second voyage 'veiled in obscurity and error' – and how the explorer confiscated sea charts that his crew had made. Thus the setting of Sauer's story remains hazy, like the 'mixture of fact, fancy, and credulity' (Sauer 1966: 23), replete with cannibals and other imagined dangers, in Columbus's mind. Like adventure heroes, who are continually tormented and tested, Sauer's Columbus believes himself to be tested by his creator:

> Columbus cited Biblical and Apocryphal writ, and a trance in which a voice assured him that all his tribulations were inscribed in marble and that his Creator was testing him for the reward he would have.
>
> (Sauer 1966: 136)

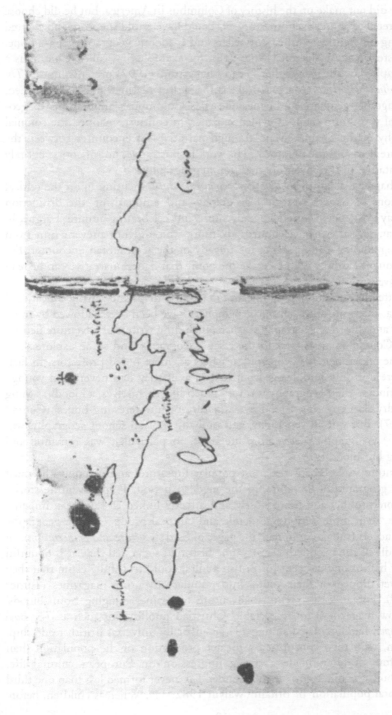

Plate 1.4 The Northwest Coast of Española, a sketch attributed to Columbus, which was reproduced in Sauer's *The Early Spanish Main* (1966: 40). The sketchy but realistic point of departure for Sauer's academic, geographical adventure story.

Sauer did not make up the history of Columbus in America, but he did choose how to retell it: as an adventure story. He could have retold the story and charted its setting differently, as other storytellers and academic geographers have done. Map historian R.A. Skelton (1968: 218), for example, has called Columbus a 'cartographer', while geographer David Livingstone (1992) presents him in *The Geographical Tradition* as a practical man, a skilful mariner and accomplished, knowledgeable geographer. Livingstone cites Columbus's annotations to geographical works and maps as evidence. Sauer, in contrast, mapped the colonial geography of the Spanish Main through narratives of encounter between the adventure hero and the unknown. His work is illustrative, not of course entirely representative, of the adventure tradition in geography.

In Chapter 8 I return to *The Early Spanish Main*, focusing upon the critical dimensions of Sauer's historical geography, and emphasising the distinction (that may exist) between retelling adventure and celebrating empire. I argue, in particular, that if Sauer celebrated adventure, this did not prevent him from criticising empire – from seeing the tragedy of the Columbian encounter. For now, though, I simply wish to make the point that Sauer's narrative can be read as an adventure story.

Despite their undoubted significance in making maps and setting educational curricula, academic and scientific geographers have been overshadowed by more popular and informal geography teachers, including writers of adventure fiction. Joseph Conrad suggests why in the essay 'Geography and some explorers', in which he remembers his geography teachers as 'mere bored professors, in fact, who were not only middle-aged but looked as if they had never been young'. He continues, 'their geography was very much like themselves, a bloodless thing with a dry skin covering a repulsive armature of uninteresting bones' (Conrad 1926: 17). But not all geography, and probably not all school geography, was so dull.[14] Geographical adventure literature, in particular, was romantic and anything but bloodless.

Adventure was perhaps the most popular literature, and certainly the most popular geographical literature, of the modern period. Pulp adventure stories were more prolific, and more closely and directly linked to European imperial outreach, than more 'serious' academic and cultural products, from geography texts to novels (such as dominate the pages of Said's *Culture and Imperialism*). In the middle of the twentieth century, one historian suggested that 'it is doubtful whether half of one per cent [of British adults] could truthfully claim that they had never dipped in their youth' into popular adventure magazines (Turner 1957: 9). For at least a century adventure magazines, ranging from 'bloods' (low-brow magazines) to 'respectable' Christian publications such as *The Boy's Own Paper*, had circulated among a large, virtually universal British readership. Children, who then constituted a greater proportion of the population than ever before or after, made a strong impression on European cultural life. Throughout the nineteenth century, under-14s 'never formed less than one third of the total population' of Britain (Walvin 1982: 11). Whereas children, before and after, have tended to read books and magazines designed mainly for adults, the opposite was true for a time, particularly between the middle of the nineteenth and twentieth centuries.

Adventures, although often marketed as boys' stories, attracted readers of both sexes, and (almost) all ages and classes. Boys, it has been found, generally read more adventure stories than their female counterparts, but many of the other genres traditionally favoured by girls, notably school and detective stories, contain elements of adventure, even though they are not defined principally as adventure stories (Jenkinson 1946: 174). And the popularity of European adventures was not limited to Europe; they were consumed around the world. Surveys show, for example, that *Treasure Island* was the most popular book among American boys in 1926, and also among New Zealand boys as late as 1947.[15] In 1974 the nineteenth-century French adventure story writer Jules Verne, and his twentieth-century British counterpart (in terms of popularity, at least) Enid Blyton, were the world's most widely circulated – the most translated – authors, after Marx and Lenin (Ray 1982). Blyton's worldwide sales exceeded 500 million books, including over 20 million of the *Famous Five* adventure series (Mullan 1987). Only recently have British adventure story writers like Bessie Marchant and Robert Ballantyne, household names around much of the English-speaking world until after the Second World War, finally faded into obscurity.[16]

Popular adventure literature (in Europe), and popular literature in general, was a largely nineteenth- and twentieth-century phenomenon. Before the nineteenth century, relatively few people could afford to buy books or could read, although adventure literature was popular by contemporary standards. *Robinson Crusoe*, appealing to many people who were not particularly 'literary', but who read mainly for pleasure, became one of the best-known books in early eighteenth-century Europe (Baker 1931). *Robinson Crusoe*, it has been said, 'not only created a new literary form; it created a new reading public' (Moore 1958: 222). But Defoe did not, of course, create a new reading public single-handed. The growth of the reading public was a response to social and economic changes that enabled more Europeans to read and to afford books and magazines. In Britain in 1840, around two-thirds of men and one-half of women could read, and these proportions increased to over nine-tenths of men and three-quarters of women by the end of the century, as more and more working-class people learned to read. By the outbreak of the First World War, illiteracy was virtually wiped out.[17] Throughout the nineteenth century, and into the twentieth, adventure books and magazines became more accessible to popular audiences, cheaper to buy, easier to read, more appealing to look at. This was partly a result of technical and industrial changes in the nineteenth century that made it possible to produce books that were brighter and more heavily illustrated, and to do so cheaply. In the period between 1860 and 1880, in particular, there was 'a flowering of illustration' in British books and magazines, as reproduction techniques including colour lithography and colour printing from wood became more commercially viable (Whalley and Chester 1988: 75). Visually appealing, cheap and generally accessible, European popular adventure literature reached very broad audiences in the nineteenth century. Its impact on European culture, and on European geographical imaginations, was huge.

GEOGRAPHICAL IMAGINATION AND ADVENTURE

Geographical imaginations make possible, but do not determine, particular social and cultural politics. Geographical imagination has been defined, quite openly, as 'sensitivity towards the significance of place, space and landscape in the constitution and conduct of social life' (Gregory 1994a: 217). To think about social and cultural life geographically is, potentially, to entertain thoughts that are otherwise inconceivable. It is to change the way society and culture are conceived. In social and cultural studies, this means reasserting the role of space in critical social theory (Soja 1989), and asserting that culture is spatially constituted (Jackson 1989). In *Geographical Imaginations* (1994b), geographer Derek Gregory emphasises the fluidity and, as his title suggests, the plurality of geographical thought, of mental ground on which thoughts rest. The mental ground upon which people think about society and culture affects and limits what they think. It has been suggested, for example, that patriarchal capitalism was founded upon conceptions of absolute space that have been adopted unproblematically by Anglo-American human geographers (Katz and Smith 1993). To free geographical social thought from absolute space may, it follows, be to free those thoughts from patriarchal capitalism. Frederick Jameson (1988), similarly, emphasises the need to create new conceptual space, in what he calls 'cognitive maps'. In the post-modern condition, he suggests, we have lost a sense of the social totality. Only through a new 'aesthetics' of cognitive mapping, in which the social totality is represented, made imaginatively accessible to the people, will a politics of that totality become possible. Cognitive mapping, a geographical metaphor, for which Jameson is indebted to humanistic city planner Kevin Lynch, is both spatial and concrete, referring to cultural *space* which may be accommodating to political action (Phillips 1993). But, as Jameson also argues, to create cultural space in which new Socialist politics are possible is not necessarily to spawn such politics, nor to determine their form. The cognitive map, which Jameson regards as a precondition for new Socialist politics, may become a vehicle for other, non-Socialist politics. The space he envisages is therefore open-ended.

Adventure stories constructed a concrete (rather than purely abstract) cultural space that, like the cognitive maps advocated by Jameson, mapped a social totality in a manner that was imaginatively accessible and appealing to the people. And like Jameson's cognitive mapping, the construction of geographies of adventure was generally – but not universally – motivated by a clear political agenda: broadly speaking, imperialism. Adventure stories constructed cultural space in which imperial geographies and imperial masculinities were conceived. But, as Jameson argues, the intentions of writers and other map-makers, be they imperial or Socialist, need not determine the ways in which their maps work. Once in cultural circulation, their maps may have unintended consequences. So it is that geographies of adventure, in cultural circulation, make possible politics and political actions that their creators never intended. In particular, geographies of adventure accommodate politics of resistance to dominant constructions of imperialism and masculinity.

The geography of adventure, a cultural space opened up by European encounters with the non-European world and by European narratives of encounter with the non-European world, introduced a new dynamic to both places. Adventure has had profound implications for the non-European world – the material space in which stories are set – but also for Europe – the often-invisible home of the adventurer (home residence, area and country). Superficially, adventure stories say little about home, since they seem to focus on what Joseph Campbell calls the 'zone unknown' of adventure.

> This fateful region of both treasure and danger may be variously rep-resented: as a distant land, a forest, a kingdom underground, beneath the waves, or above the sky, a secret island, lofty mountaintop, or profound dream state; but it is always a place of strangely fluid and polymorphous beings, unimaginable torments, superhuman deeds, and impossible delight.
>
> (Campbell 1949: 58)

The settings of modern adventure, likewise, are said to be divided between home and away. Green (1979: 23), for example, argues that 'adventure seems to mean a series of events, partly but not wholly accidental, in settings remote from the domestic and probably from the civilised'. The two distant, disconnected spaces, divided by perilous and extraordinary voyages, seem to have little to do with each other. And yet they are defined in terms of each other, and the boundaries between them are often blurred (Dawson 1994). Unknown, distant spaces of adventure are vehicles for reflecting upon and (re)defining domestic, 'civilised' places. With elements of a recognisable world, recombined in a different order and located somewhere off the edge of the map, on or around the margins of the known world, the geography of adventure corresponds to what drama critic Victor Turner (1969, 1982) calls a liminal space.[18] It is a marginal, ambiguous region in which elements of normal life are inverted and contradictions displayed. Turner argues that liminal spaces tend to be conservative in tribal societies, where they accommodate traditional rites of passage and therefore reinscribe the social order, and subversive in complex, industrial societies, where they disturb the social order. In industrial societies, liminal (more precisely, liminoid) spaces tend to be associated with leisure, for example in the material landscapes of beaches and theme parks (Shields 1991; Warren 1993) and the metaphorical landscapes of popular adventure stories, including those in print.[19] In the liminal geography of adventure, the hero encounters a topsy-turvy reflection of home, in which constructions of home and away are temporarily disrupted, before being reinscribed or reordered, in either case reconstituted.

13

MAPPING ADVENTURES

The geography of adventure is cultural *space* in which identities and geographies are constructed, and its spatiality is reflected in those constructions. Identities and geographies reflect the *geography* of adventure; they are *mapped*. But what does it mean to say that adventure stories *mapped* geographies and identities? The language of mapping, both literal and metaphorical, is perhaps more prolific than it is precise. Among social and cultural critics, in particular, 'these are boom years for cartographic metaphors; theories, projects, concepts and differences are all being mapped' (Katz and Smith 1993: 70). Used differently in different contexts, cartographic and spatial metaphors are both slippery and fluid. They are contested terms, unstable, uncircumscribed, and therefore continually able to open new conceptual spaces, in which new forms of social and political action may be conceived. So it would be counter-productive to begin with a formal definition of mapping. But, if it is left completely open, mapping is too slippery to be conceptually – rather than just rhetorically – useful. What, then, does mapping mean, and when is an adventure story a map?

Maps *may* be spatial, visual, graphic representations, but the information they represent *must* be spatial. The cartographic map, in which 'space is used to represent space' (Andrews 1989: 16), is a special form of map; it is not the only form of map. Cartographers do not have a monopoly over maps. Indeed, as map critic J.B. Harley (1992: 231) put it, 'Maps are too important to be left to cartographers alone.' Maps are broader, also, than their entry in the *Oxford Dictionary* suggests: 'map n. a usu. flat representation of the earth's surface or part of it' (Thompson 1995: 831). Maps need not describe material space; they may, for example, represent geographies of fantasy (Flood 1993; Lewis 1985) and imagination (Wright 1947; Kuipers 1982). While maps are distinctive in the sense that they describe material and metaphorical spaces, they are otherwise similar to other forms of representation. Art historian Svetlana Alpers (1983) argues that there are no intrinsic differences between maps and other forms of descriptive picture, despite commonly held assumptions that maps (not pictures) are authoritative and truthful. Harley (1992) conceptualises the relationship between maps and other forms of geographical description by showing how both can be read as cultural texts. Cartography, he argues, is geographical discourse. Given this broad definition of maps, the adventure story, replete with vivid geographical imagery, is as much a map as any formal cartographic image.

Adventure stories share with other maps, particularly cartographic maps, a measure of authority, a power to naturalise constructions of geography and identity. Maps possess what Alpers (1993: 133) calls the 'aura of knowledge'. They are commonly regarded as scientific, objective and mechanical. It seems that 'they depict the world as it "really is"; as the author vanishes in the map, the map exudes authority' (Smith 1994: 499). The authority of maps lies in their ability to circumscribe geography, by enclosing, defining, coding, orienting, structuring and controlling space. It also lies in their propensity to ignore, suppress and negate alternative geographical imaginations (Huggan 1994). Thus cartographic representations are naturalised as facts, at least in the popular geographical imagination. Intellectuals, demonstrating the possibility of deconstructing maps by 'read[ing] between the lines of the map' (Harley 1992: 233), are perhaps the exception that proves the rule; most people do not deconstruct maps, they accept them as fact. The taken-for-granted world of the map naturalises ways of seeing, ways of reading the landscape. It also naturalises the social relations embedded within those ways of seeing (Duncan and Duncan 1988).

Like other modern maps, realistic adventure stories naturalise the geographies they represent, and normalise the constructions of race, gender, class and empire those geographies inscribe. The realistic, plain language of *Robinson Crusoe*, for example, seems to report the 'truth' about adventurers and their settings. Defoe's language is as plain, as spare, as naturalistic and even as mechanical in appearance as (what Harley calls) cartographic discourse. The geography of *Robinson Crusoe* is constructed in plain English, not the sentimental prose one might expect from a popular adventure story. As Virginia Woolf (1932: 54) once commented, 'There are no sunsets and no sunrises [in *Robinson Crusoe*]; there is no solitude and no soul. There is, on the contrary, staring us full in the face nothing but a large earthenware pot.' *Robinson Crusoe* is replete with realistic images of everyday objects (like earthenware pots) and realistic images of geography. Defoe's geographic realism is illustrated, reiterated and reinforced in the map illustration, which first appeared in the fourth edition (dated 7 August 1719) of *Robinson Crusoe*. In the 'MAP of the WORLD, on wch. is Delineated the Voyages of *ROBINSON CRUSO*' (Plate 1.5), longitude and latitude are labelled, geographical knowledge clearly marked, while geographical ignorance is made plain, according to the minimalist cartographic aesthetics and conventions of the day. Looking like a map rather than a 'mere' picture, and reflecting contemporary geographical literature and atlases, with which Defoe was familiar (Baker 1931), the image assumes some intellectual authority. However, Defoe admitted in the preface that his realism was superficial: 'The Editor believes the thing to be a just History of Fact; neither is there any Appearance of Fiction in it' (Defoe 1719a: ii).[20] Upon close inspection, the realistic appearance of the map, for example, proves superficial, its place names, latitude and longitude markings barely legible and quite useless. But the general appearance of fact, rather than the subtle disclaimer buried in the preface, led many readers to believe they were reading a 'real' map and a 'true' story, in which geographical 'facts' were faithfully represented. Superficially, at least, the geography of adventure is unambiguously realistic, an uncomplicated world of social, moral and political certainty. And superficially, at least, the geography of adventure is solid ground, on which adventurous imaginations and colonial adventurers are free to move.

When told and retold many times according to conventional formulae, adventures tend to reinscribe received geographies and identities, and they therefore tend to be conservative. Few stories, for example, have been more

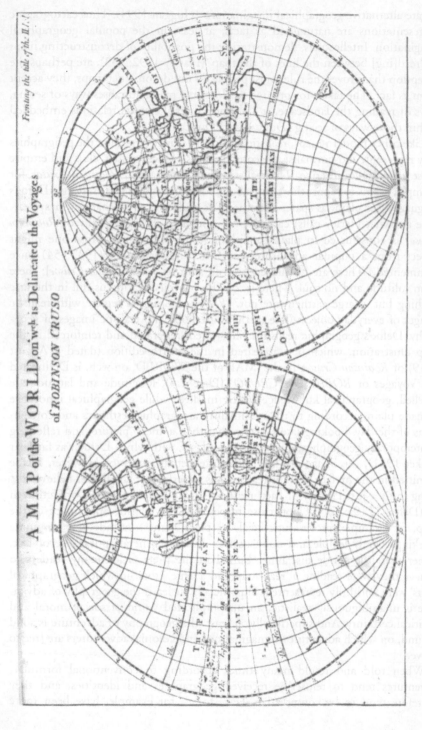

Plate 1.5 'A MAP of the WORLD, on wch. is Delineated the Voyages of *ROBINSON CRUSO*, dated 7 August 1719. This illustrates the appearance of reality in the geography of adventure.

conservative, more naively realistic (and politically loaded), than *Robinson Crusoe*, as it was retold and imitated in nineteenth-century Britain. Nineteenth-century Robinsons and Robinsonades mapped Britain, on the one hand, and the British Empire, on the other, as I argue in Chapter 2.[21] The realistic geography of the Robinsons and Robinsonades was an imaginary space in which constructions of Britain and empire were naturalised. Robinson Crusoe's island was a middle-class, Christian, British man's utopia. *Robinson Crusoe* and other adventure stories mapped many aspects of Britain ('home') in relation to the island. They mapped a world view that placed Britain at the (imperial) centre and colonies like Crusoe's island at the margins. They mapped British constructions of race (roughly speaking, white Crusoe in relation to non-white 'savages'), its class system (Crusoe as master, Friday as slave), its gender (Crusoe as masculine, nature as feminine), religion (Crusoe as Christian, 'savages' as non-Christians) and language (Crusoe has spoken and written command of the English language, Friday is relatively mute). These constructs are interrelated, and cannot really be separated. It is possible, though, to focus on individual aspects of the ways in which adventure stories mapped home (Britain) and setting (including British colonies), as I do in Chapters 3, 4 and 5.

Refocusing upon particular processes of geographical and social mapping means turning from world-famous but geographically or socially generic narratives like *Robinson Crusoe* to works of more specific geographic and social importance. Focusing upon specific, contextual processes of mapping means turning from the general to the particular, from the canonised to the relatively obscure, from stories that are often considered important in their own right to stories that are important and suggestive locally or as vignettes. It also means turning from global geographies and whole empires to particular colonies.

Canada and Australia[22] provide opportunities to explore the fullest possible range of material and metaphorical mappings and unmappings. I select stories set in Canada and Australia because those colonies, which started out as British settler societies, engaged popular audiences at a wide variety of levels, ranging from the purely imaginary to the more practical. Unlike stories set in the Brazilian rainforest, among African gorillas and in Antarctica, Canadian and Australian stories constructed settings that most readers could reasonably imagine going to, and that many readers did go to. British settlers also continued to read and write, produce and consume, adventure stories set in Canada and Australia, after their arrival in those colonies. The authors and stories that I have selected to read in detail, in Chapters 3 to 5, are important and representative in their geographic and social contexts. Robert Ballantyne (Chapter 3) was prominent among adventure writers who set stories in western Canada in the Victorian and Edwardian period, while Ernest Favenc (Chapter 4) was among the best-known and most influential writers to set adventures in Australia in the same colonial era. Bessie Marchant (Chapter 5) was important mainly as a pioneer of girls' adventure stories, about and for girls, and as a champion of female emigration to colonies and dominions including Canada.

In Chapter 3 I focus on the mapping of British gender in the realistic spaces of adventure, specifically an adventure by R.M. Ballantyne – *The Young Fur Traders* (1856). Ballantyne charted imaginative geography in which he mapped masculinity. He was commended by parents, teachers and critics for educating his readers in geography (Salmon 1888: 58), but his realistic geography was not a static or neutral classroom concept. It was space in which he articulated an ideology of masculinity, specifically a variation on the mid-century construction of masculinity known as Christian manliness (see Vance 1985). While adventure is sometimes labelled a 'masculinist' narrative (for example by Green 1991), adventures do not reinscribe archetypal, singular masculinity; they map historical masculinities. Mapping in this sense is metaphorical rather than literal, since literal maps describe geography rather than identity. The metaphor is meaningful in this instance because masculinities are spatially constituted; they reflect the characteristics of the spaces in which they are constructed. Masculinities mapped in the geography of adventure reflect the characteristics of that geography.

Colonies, like masculinities, are mapped in the geography of adventure. They are conceived in spaces of adventure and reflect the characteristics of those spaces. While *The Young Fur Traders* can be read in relation to (conservative) gender politics in Victorian Britain, it can also be read in relation to (conservative) imperial politics in Britain and British North America. Ballantyne's novel was set in a region governed by the London-based Hudson's Bay Company (HBC), and it helped describe that region to readers in Britain and around the world. Adventure stories like *The Young Fur Traders* imaginatively mapped imperial geographies, and they also inspired merchants, investors, travellers, settlers and others to go out and physically become 'empire builders'. In the following two chapters I show how adventure stories mapped particular imperial spaces, how writers and readers of adventure stories imaginatively and physically mapped and made particular imperial geographies. I place adventure stories in their Victorian (historical, geographical and political) contexts, but here the context is mainly colonial – Australia and western Canada – rather than British. In context, the adventure stories are interpreted as a part of the process of colonisation, of making imperial geographies. Ernest Favenc's exploration adventure stories (Chapter 4) mapped an unknown and mysterious central Australian space in which colonists could imagine an Australian nation. This colonial geography, in stories such as *The Secret of the Australian Desert* (Favenc 1896a), bears traces of the geography of adventure, particularly its masculinism. Favenc's masculinist Australian geography was colonised, not physically but imaginatively, as ideological territory. So long as it retained some of its mystery, and remained somewhat unmapped and unknown, central Australia remained a space in which anything seemed possible.

Other adventure story writers also mapped colonial spaces, but rather than mapping unknown spaces with indeterminate, open-ended colonial futures, they mapped with immediate, specific colonisation projects in mind (Chapter 5). Writers such as James Macdonald Oxley in Canada and Bessie Marchant in Britain

18

used adventures to promote emigration to, and settlement in, Canada. Oxley ameliorated the settings of adventure stories in order to represent Canada as a destination for emigrants and settlers, although he did not actually write settlement novels, as Marchant did. Marchant's emigration and settlement novels mapped Canada as a destination for female emigrants, who would become settlers.

UNMAPPING ADVENTURES

Although maps, from road atlases to adventure stories, are generally 'authoritarian images' that 'reinforce and legitimate the *status quo*' (Harley 1992: 247), they also open conceptual space for more critical politics. As a text, the map is inherently unstable, its meaning open to challenge and change. The meaning(s) of the map text, not determined by the intentional author (cartographers are often anonymous, anyhow), and neither fixed nor singular, is socially and historically constructed (Duncan and Duncan 1988). The changing meaning of *Robinson Crusoe*, for example, illustrates the adaptability of adventure stories and spaces (Chapters 2 and 7). The space that adventures put into cultural circulation may be appropriated as a vehicle of authority, reinforcing hegemonic masculinity and imperialism, for example. But the geography of adventure may be a point of departure for critical politics, like the enabling, 'wayfinding' maps envisaged by geographers Steve Pile and Nigel Thrift (1995: 1), and the malleable, slippery maps of Deleuze and Guattari. They write,

> The map is open and connectable in all of its dimensions; it is detachable, reversible, susceptible to constant modification. It can be torn, reversed, adapted to any kind of mounting, reworked by individual group or social formation.
>
> (Deleuze and Guattari 1988: 12)

The geography of adventure can be as slippery as the maps of Deleuze and Guattari, a point of departure for criticism and resistance.

In Marchant's stories, maps provide girls with points of departure. Heroines proceed to transgress some of the boundaries on those maps. They cross literal boundaries, between home and away, and metaphorical boundaries, between constructions of femininity and masculinity. While Marchant mapped colonial acts and colonial geographies, in many ways as conventionally and conservatively as writers such as Ballantyne and Favenc, she introduced a note of ambivalence to the adventure story. Her colonial maps challenge as well as confirm dominant ideologies of gender, imperialism and geography. Bessie Marchant illustrates an ambivalence that is present, to greater or lesser extents, in all adventure stories. On the one hand, Marchant's heroines challenge male domination of adventure literature (heroes, readers), and challenge male domination of travel and emigration. On the other, Marchant's girls are conventional colonists, on the side of the British colonial establishment, planting British society in North America. On the one hand they are drawn to a space of adventure, on the other they challenge that

19

space, attempt to subdue it and transform it. They are both rebellious emigrants and sedentary settlers. Thus Marchant's adventure stories were conservative, mapping colonial geographies in conventional ways, but, at the same time they were radical departures, unmapping rather than mapping. In Marchant's girls' stories, as in a wider range of feminist fictions, 'the clear lines on old patriarchal maps become blurred or effaced', as Graham Huggan has put it (1989b: 18). Through their transgressions, Marchant heroines undermine or unmap some aspects of the predominantly masculinist geography of adventure.

The ambivalence and fluidity of adventure's geography – authoritarian in some readings, subversive in others – is illustrated particularly well in extraordinary voyages by Jules Verne and others, as I argue in Chapter 6. A tradition of extraordinary voyages can be identified that stretches back from Verne in the late nineteenth century, through Jonathan Swift and Daniel Defoe in the early eighteenth century, and Gabriel de Foigny in the late seventeenth. These voyages, despite strong similarities, have been read very differently, some labelled conservative, others critical. The difference lies more in the reading than in the stories, and this illustrates the power of certain readers to determine the meaning of the texts. I compare readings of Foigny and Verne. Opinion leaders, biographers and literary critics have generally labelled Foigny a radical and Verne a conservative. Dominant readings of Gabriel de Foigny's influential extraordinary voyage to Australia illustrate the possibility of reading adventure as revolutionary social criticism, and reading adventure's geography as space in which such politics are conceivable. Foigny's narrative, of a perilous journey that ends in Australia, may be read as an allegory of the upheaval France would have to go through before it could reach some vaguely utopian state. In contrast, Verne's extraordinary voyages have been read, particularly in Britain, as pro-imperial, conservatively masculinist narratives. Alternative readings of Verne, as an anarchist and an anti-imperialist, an enemy rather than a servant of the British Empire, have largely remained latent until after the Second World War. Verne's geography, although appropriated and charted by dominantly conservative readers, was not intrinsically conservative.

The geography of adventure has been reused as a site of resistance by a range of post-colonial critics, who unmap (literally) the geographies of empire, and (metaphorically) the identities, particularly the imperial masculinities, constructed in that geography. I explore unmapping, like mapping, mainly with reference to selected stories, some of which are read and known locally rather than internationally, and are illustrative genre novels rather than world-famous or canonised texts. Although some adventures map masculinity and imperialism, adventures are neither intrinsically 'masculinist' nor 'imperialist'.[23] Girls' adventure stories, emphasising forms of womanliness, and adventure stories set in Britain rather than empire, make nonsense of such sweeping claims (although girl heroes can be constructed as masculine, and British settings as colonial). Rather than necessarily confirming entrenched images of masculinity and imperial encounter, adventure stories can explore those images, and sometimes contest them. In Chapter 7 I show how some post-war writers, including William Golding, J.M. Coetzee and Michel Tournier, have re-entered the spaces of adventure, in order to contest and reinvent some of the masculinist, racist and imperialist constructions of earlier adventure stories, including *Robinson Crusoe* and some Robinsonades. Golding unmapped British imperial masculinity in the geography of adventure, geography charted originally by writers such as Ballantyne. Coetzee and Tournier unmapped the geography of adventure itself, and therefore undermined the constructions of masculinity and imperialism that were naturalised within it. Continually being mapped and unmapped, the geography of adventure, like the space of the map, is more fluid and adaptable than is sometimes apparent.

(RE)READING ADVENTURES: WHAT FOLLOWS

We should gloat over a book, be wrapt clear of ourselves, and rise from the perusal, our mind filled with the busiest, kaleidoscopic dance of images, incapable of sleep or of continuous thought.
(Robert Louis Stevenson, cited by Salmon 1888: 105)

Robert Louis Stevenson argued that readers should allow themselves to become absorbed in the worlds of adventure, intoxicated by the brightly coloured, almost magical images of distant lands and seas. Similarly, Ballantyne's Ralph Rover warned readers of *The Coral Island* that they must 'enter with kindly sympathy into the regions of fun', and advised 'any boy or man who loves to be melancholy and morose' to 'shut my book and put it away. It is not meant for him' (Ballantyne 1858a: iii). Stevenson and Ballantyne suggested, in effect, that to understand adventure one must allow oneself to be seduced by it, and deny oneself critical, intellectual distance. I do not wish to pour cold water onto Ralph's regions of fun, nor to be disrespectful of those who have willingly followed him there and fantasised about adventures of their own in 'little-known' parts of the earth. Nor do I wish to deny that I, as a child and perhaps as an adult, have been entertained by adventure stories, and seduced by dreams of escaping to other worlds, and the possibility of finding adventures – and the violence and liberation they promise – there. But, without completely detaching myself from this geographical fantasy, I stand back from it, just far enough to interpret adventure stories, to think about what they mean and how they function as popular, geographical narratives. In order to stay close to the stories themselves, I write in the present tense as much as possible, and pay particular attention to plot, since these are action stories, defined more by what happens than by anything else. I assume greatest analytical distance when interpreting settings, the spaces opened up by tales of adventure. Reading a series of adventure stories, I begin to explore the spaces of adventure, and the ways in which identities and geographies have been, and continue to be, mapped and unmapped within those spaces.

2

MAPPING ADVENTURES
Robinson Crusoe and some Victorian
Robinsonades

Among the books of travel and discovery published in the modern period, none
has made a greater impression on geographical imaginations than *Robinson
Crusoe*, the single most famous, representative and influential adventure story of
the time. Popular since its publication in 1719, Daniel Defoe's original novel has
since been transformed and redefined, with many different editions, abridgements, imitations and readings. The production of 'Robinsons' was most prolific
in the nineteenth century when the story took its place among the foundational
myths of British culture. It held that place until after the Second World War,
when one critic argued that 'Almost universally known, almost universally
thought of as least half real, [*Robinson Crusoe*] cannot be refused the status of
myth' (Watt 1951: 96).

Robinson Crusoe illustrates and introduces the broad outlines of the geography
of adventure, and the ways in which identity and territory are mapped within
them. The story maps British identities in its setting, mainly the island where
Crusoe is shipwrecked, and it maps British colonies in the same space. This
process of mapping involves naturalising and normalising constructions of
identity and geography which, in the nineteenth-century context, were conventional and conservative. A geographical medium, naturalising and normalising
conservative constructions and ways of seeing, *Robinson Crusoe* can be regarded
as a map, both metaphorically and literally, of the British Victorian world.
I begin this chapter by reading an abridgement of *Robinson Crusoe*, typically
undated and broadly representative of the novel as most nineteenth-century
British readers knew it.

An extremely influential book, *Robinson Crusoe* was also much copied, in
books known as Robinsonades. Robinsonades mapped variations on the generic
theme of British identity and imperial geography. I read two mid-nineteenth-
century Robinsonades, *The Coral Island* (1858a) by Robert Ballantyne and
Canadian Crusoes: a Tale of the Rice Lake Plains (1852) by Catharine Parr Traill,
to illustrate the ways in which specific Robinsonades mapped British Victorian
world views and particular British colonies.

22

ROBINSON CRUSOE IN NINETEENTH-CENTURY BRITAIN

The Life and Strange Surprizing Adventures of Robinson Crusoe, of York, Mariner (Plate 2.1), published early in the eighteenth century, was in many ways a book of its time. It was the work of Daniel Defoe, a journalist who addressed contemporary political concerns from the perspective of a Whig activist, Dissenter, spy and merchant. Defoe's language and his images in *Robinson Crusoe* were also very contemporary. His narrative form was indebted to the eighteenth-century French tradition of imaginary or extraordinary voyages (Atkinson 1966),[24] and more specifically to the few Robinsonades, island adventure/utopias such as Henry Neville's *Isle of Pines* (1668), which were already in print.[25] Defoe's central image, that of a castaway struggling to survive on a desert island, was borrowed from a shipwreck story, published in 1709, which described the adventures of Alexander Selkirk, a castaway on the South Pacific Juan Fernandez islands.[26] Defoe's geographical and anthropological images, also, were adapted from early eighteenth-century literature. Robinson Crusoe's voyages took place in territory charted by contemporary travel writers, explorers and cartographers (Plate 1.5), although Defoe did use geography 'for his own ends', taking 'liberties' with it 'when the necessities of the story so demanded' (Baker 1931: 257).[27] His story, early eighteenth-century both in its politics and in its images, made an immediate impression on contemporary readers and critics, enjoying great success among a readership of unprecedented breadth.

Despite its immediate success, *Robinson Crusoe* was not to make its greatest impact, and its greatest impression upon geographical imaginations, until the nineteenth century. Published when 'there was no English novel worth the name, and no book (except the Bible) widely accepted among all classes of English and Scottish readers' (Moore 1958: 222), the impact of *Robinson Crusoe* in eighteenth-century Britain should not be underestimated. But to suggest, as one critic has, that the book 'not only created a new literary form; it created a new reading public' (Moore 1958: 222), is to overstate the case. The early eighteenth-century readership was broad, indeed, but only by contemporary standards. Readers were concentrated in the small book-buying and book-reading public, comprised largely of the urban middle class (Watt 1957). Not until the nineteenth century was 'popular literature' truly popular across most of the geographical and social spectrum.[28] While forty-one editions of *Robinson Crusoe* were published in Britain within forty years of its publication, the total had risen to at least 200 by the end of the nineteenth century. The production of Robinsons peaked in the Victorian period, with an average of more than two per year. In addition, 110 translations appeared in print before 1900, alongside at least 115 revisions (Shinagel 1975; Green 1990). These assumed every conceivable form, ranging from *Robinson Crusoe in Verse* (Bott 1882) to *Robinson Crusoe in Words of One Syllable* (Godolphin 1868). Godolphin replaced Bible, for example, with the monosyllabic 'Book of God's Word' (Godolphin 1868: 29),

Plate 2.1 Original frontispiece of *Robinson Crusoe* (1719). Despite hundreds of editions, abridgements and imitations, John Pine's original engraving remains the most famous image of Robinson Crusoe.

and after the title she made only two exceptions to the one-syllable rule.[29] Typical of many nineteenth-century Robinsons, *Robinson Crusoe in Words of One Syllable* was written in simple language, illustrated with crude colour illustrations, and aimed at very young children (Plate 2.2). A survival story and spiritual biography, the *Robinson Crusoe* story was canonised as the archetypal modern adventure story, and as a foundational myth of modern, enlightened, imperial Europe. In the eighteenth century, Jean Jacques Rousseau (1762) became one of the first to endorse Defoe's story, which he saw as 'a complete treatise on natural education' that would 'serve as our guide during our progress to a state of reason' (Rousseau 1762: 2.64). Many famous (and not-so-famous) writers and literary critics – including Karl Marx, John Ballantyne (Robert's uncle), Samuel Taylor Coleridge, James Joyce and Virginia Woolf – followed Rousseau, reading new meanings into *Robinson Crusoe* and helping to canonise it as a 'great' literary work, perhaps as a myth.[30]

The geography of *Robinson Crusoe*, unsettling and even radical to many eighteenth-century readers, became familiar, accepted by nineteenth-century readers as 'real'. In its politics and style, *Robinson Crusoe* was a departure from the early eighteenth-century mainstream, a radically different kind of story. Defoe, an accomplished writer and geographer, was not immediately adopted by the contemporary literary and geographical fraternities,[31] neither of which knew what to make of his plain, popular style or his realistic fiction (Rogers 1979; Seidel 1991). A plainly written narrative by a man whose 400 or so previous publications were generally regarded as journalistic scribbles, *Robinson Crusoe* did not seem like serious literature. Neither did it seem like serious geography. Defoe was, however, a geographer – in the threefold sense that he was familiar with contemporary geographical literature, was a seasoned traveller, and was a writer of geographical narratives and descriptions. He 'took immense pride in his knowledge of geography; his library was well stocked with atlases and works on discovery and navigation; he was forever surrounded by maps and charts' (Rogers 1979: 25). Defoe's geography was informed by, and written in the language of, these contemporary geographies. But Defoe was a storyteller rather than a chronicler, a travel writer rather than a traveller, as he hinted in the introduction to *A New Voyage Round the World*. 'A seaman when he comes to the Press,' he reflected, 'is pretty much out of his Element, and a very good Sailor may make but an indifferent Author' (Defoe 1725: 1). Contemporary critics such as fellow-journalist Charles Gildon accused Defoe of deception, arguing that his fiction was sometimes mistaken for fact (Gildon 1719). Geographers, unable to see the truth in fiction, have often repeated the charge, and expressed distaste for what one RGS Secretary described as 'stories of travel which are actually meant to deceive, which are pure fiction from beginning to end, passed off as fact' (Keltie 1907: 186).[32] Typically, when geographers read Defoe (particularly Defoe's fiction) they regard him as 'a journalist, a writer of political pamphlets, and of stories for human entertainment', and not as a geographer, since 'he did not set out to teach his readers geography' (Baker

Plate 2.2 Frontispiece and title page of Godolphin's *Robinson Crusoe in Words of One Syllable* (1868). A nineteenth-century Robinson, abridged, simplified and illustrated for children. The basic, flat colours (yellow, orange and green over black) combine to produce a stylised and generally crude appearance, which mirrors the simplicity of the text. Colour illustrations contributed to the transformation of *Robinson Crusoe* as colour printing, technically possible since the 1830s, became commercially viable in the 1860s.

1931: 257). But while many geographers continued to distrust and marginalise realistic fiction, other readers and critics came to terms with it. Theorising the novel, critics decided that Defoe's 'circumstantial realism' was not deceitful, as they previously supposed. Since readers were no longer so preoccupied with the distinction between fiction and 'fact', no longer so disturbed by the idea of realistic fiction, realistic images were no longer discredited by the label 'fiction', and it was possible for Defoe's realistic geographical images to be naturalised, to be accepted as 'real'.

Unfamiliar and disturbing to many early eighteenth-century readers, the literary form and politics of *Robinson Crusoe* were familiar and normal to their nineteenth-century counterparts. Defoe's eighteenth-century innovations became nineteenth-century conventions. *Robinson Crusoe* helped invent the British novel. Its industrious, Christian hero displayed what was later termed the 'Protestant work ethic' (Watt 1957). It 'prophesied' (as James Joyce put it) rather than chronicled British imperialism, giving some attention to relatively archaic forms of imperial adventure (slavery, mercantile and plantation capitalism), but emphasising more 'progressive' forms of colonialism. In particular, it provided a model for the mass settlement projects of the nineteenth and twentieth centuries, which were facilitated by the American Homestead Act of 1863 and its counterparts in British colonies and dominions including Canada and Australia.[33] Crusoe's conversion from risk-loving mercantile capitalist (in his wandering, slave-trading youth) to practical, sedentary farmer (in his mature years) signalled the transition of adventure from a *bourgeois* to a *petit bourgeois* ideology, and helped redefine adventure from the ideology of mercantile capitalists to what Marxist critic Michael Nerlich calls 'the ideology of the middle class' (Nerlich 1987: 263).[34] *Robinson Crusoe* seemed a very conventional, conservative narrative to many readers, partly because it was so widely copied that it had effectively established new conventions. Thus the geography of *Robinson Crusoe*, deceitful and disturbing to readers of the eighteenth century, was realistic and normal to those of the nineteenth, when the story mapped – naturalised and normalised – conservative constructions of identity and geography.

With the publication of many pirated, edited, abridged, imitated and otherwise modified *Robinson Crusoes*, there is no 'truly definitive' version of the story once told by Defoe (Watt 1951). Copyright owners were 'unwilling or unable to discourage the abridgement' of a book that had, in effect, rapidly become 'public property' (Rogers 1979: 10). Since shortly after *Robinson Crusoe* was first published, relatively few readers have encountered Defoe's complete original work, and fewer still have seen the two sequels he wrote (*The Farther Adventures of Robinson Crusoe*, 1719, and *Serious Reflections During the Life and Surprising Adventures of Robinson Crusoe*, 1720). Most who have read *Robinson Crusoe* have read a one- or two-hundred-page abridgement of some description. Many have read a children's, perhaps a boy's edition, shortened and simplified for the juvenile market, typically undated and anonymous, attributed neither to Defoe nor to the editor (who abridged and/or adapted the story).[35] Comparison between the opening sentences in the original, and in typically abridged and retold versions, illustrates the ruthlessness and creativity of many publishers.

I Was born in the year 1632, in the City of *York*, of a good Family, tho' not of that Country, my Father being a Foreigner of Bremen, who settled first at *Hull*: He got a good Estate by Merchandise, and leaving off his Trade, lived afterward at *York*, from whence he had married my Mother, whose Relations were named *Robinson*, a very good Family in that Country, and from whom I was called *Robinson Kreutznaer*; but by the usual Corruption of Words in *England*, we are now call'd, nay we call our selves, and write our Name *Crusoe*, and so my Companions always call'd me.

I had two elder Brothers, one of which was Lieutenant Collonel to an *English* Regiment of Foot in *Flanders*, formerly commanded by the famous Coll. *Lockhart*, and was killed at the Battle near *Dunkirk* against the *Spaniards*: What became of my second Brother I never know any more than my Father or Mother did know what was become of me.

Being the third Son of the Family and not bred to any Trade, my Head began to be fill'd very early with rambling Thoughts.

(Defoe 1719a: 1–2)

In comparison, the following Victorian children's edition is more modern in its language, more concise, and also more didactic, its moral colours nailed firmly to the mast.

> Robinson Crusoe was the son of a respectable merchant in Hull. The kindness of his parents had made his life very happy. His heart was not a bad one; but his love of idleness and his thoughtlessness gave his good parents a great deal of anxiety. Instead of working and learning his lessons, in order to become a clever man when he grew up, his chief pleasure was to idle about the quays. Whenever he did take a book in hand, his thoughts were always wandering away to the forest of masts in the harbour.
>
> (Defoe 1886: unpaginated)

More lighthearted and entertaining, a third version of the Robinson Crusoe story, accompanying a pantomime at the Theatre Royal, Drury Lane, begins as follows.

> Daring boys till the end of time,
> Will in every age and clime,
> Run away to sea,
> Having little notion
> Of the mighty ocean
> When it's in commotion,
> Or how sick they'll be
>
> Life on shore
> Is such a bore
> Of their schemes a fetterer –
> On the deep
> They could reap
> Wealth, and fame, *et cetera*.
>
> (Andre 1881: 2)

Although there is no definitive Crusoe, the ninety-four-page anonymous abridgement produced in London by William Darton, undated but probably published around the time Victoria became Queen, can be seen as typical.[36] The Darton edition, *The Wonderful Life and Surprising Adventures of Robinson Crusoe*, has a soft green cover and measures 10 by 17 centimetres. It falls between the dour and the slapstick, between chapbook and 'complete' editions.[37] It begins as follows.

> I was born of a good family in the city of York, where my father, who was a native of Bremen, had settled, after having got a handsome estate by merchandise. My heart began to be filled very early with rambling thoughts.
>
> (Anon. undated a: 5)

Pared down to selected essentials, *Robinson Crusoe* survives in the abbreviated form familiar to most modern readers. It is reduced to the short, simple story of a man who is shipwrecked on an island, where he learns to survive and then to prosper, where he overcomes fear, where he becomes a Christian, and where he saves and converts to Christianity a cannibal he calls Friday. Eventually, after twenty-eight years on the island, Crusoe is rescued by a passing ship.

THE GEOGRAPHY OF *ROBINSON CRUSOE*

Robinson Crusoe illustrates the general outlines of the geography of modern adventure, the dialectical geography of home and away, in which adventures are set away from home, in unknown space that is disconnected, simplified, liminal and broadly realistic.

The geography and the narrative of *Robinson Crusoe* is divided between home and away. Little is said about Robinson Crusoe's home, but everything that happens to Crusoe, everything he does and everywhere he goes, is a comment on his home – his family home and his home country – as it is and as it might be. Crusoe leaves his family home in York, against his father's wishes. Like all adventurers, he travels 'beyond the veil of the known into the unknown', as Joseph Campbell puts it (Campbell 1949: 82). The unknown space is defined in relation to the known, the unmapped in relation to the mapped. The journey begins in domestic, civilised, mapped space. But since the setting of his adventure is unmapped, it is not possible to map a course that will lead there. Crusoe must simply head for the edges of his known world and abandon himself to chance. Robinson Crusoe's first adventure begins almost the moment he leaves Britain, as 'the wind began to blow, and the sea to rise in a most terrible manner', the beginning of a terrible storm indeed (Anon. undated a: 3). It is not until later, on another sea voyage, that Defoe's adventurer encounters the storm that blows his ship 'out of our knowledge', away from 'all human commerce' and onto a sand bank near an unknown land in the Caribbean or South Atlantic (somewhere near the Spanish dominions – in the vicinity of the Orinoco – he thinks) (Anon. undated a: 21). Robinson Crusoe, along with other passengers of the stricken vessel, is attempting to row ashore when 'a wave mountain high, came rolling astern of us and took us with such fury' (Anon. undated a: 22). The wave plucks the passengers from their boat, scattering them. Like the other men, Crusoe is almost completely passive against the forces of Nature, as he admits.

> I saw the sea coming after me, as high as a great hill, and as furious as an enemy, which I had no means of strength to contend with: my business was to hold my breath, and raise myself upon the water, if I could.
>
> (Anon. undated a: 23)

Finally, the waves deposit Crusoe on an unknown shore. The setting of his adventure is completely disconnected from his home, and is completely unknown to him, and, as far as he is concerned, unnamed. The disorienting, life-threatening passage defines the island as a space that is fundamentally disconnected from the world Crusoe has known before, a space that is not on his map. The adventurer surfaces in the setting of his (principal) adventure, in space which is at first blank, unknown. 'Looking round I saw no prospect' – no prospect for himself other than 'perishing', and no visual prospect of the island (Anon. undated a: 24).

The space of Crusoe's island is a simplified, uncomplicated space, malleable in the hands of the author and his fictional adventurer. Washed up on an unknown shore, Crusoe immediately pulls himself together, and begins to explore where he is and, at the same time, who he is, what he can be. Almost immediately, he begins to look around him, seeking a 'place where I might fix my dwelling' and taking a survey of the island as a whole (Anon. undated a: 29). The island Crusoe discovers is whatever he wants it to be, and it is a vehicle for social, political and moral reflection. In the seemingly uncomplicated, simplified geography and economy of the island, Crusoe's Christian, *petit bourgeois* social outlook seems more convincing than it might have done in a more textured setting, with other people, commodity markets and landlords, for example. Ironically, perhaps, Crusoe's social development is helped rather than hindered by his isolation.

The geography of *Robinson Crusoe* is broadly realistic, plausible to nineteenth-century readers. Defoe's geographic imagery, like his hero and narrative, is suffused with the appearance of reality; it is a 'just History' without 'any Appearance of Fiction' (Defoe 1719a: ii). The original frontispiece, among the most enduring elements of Defoe's original since it is reproduced with relatively slight variation in most Crusoes, including the Darton edition, presents an image of Crusoe that reflects and reinforces the realism of the written text.[38] The image (Plate 2.1) follows Defoe's detailed descriptions, omitting only the goat-skin umbrella, saw and hatchet, which have found their way into some later illustrations of the adventurer. Like the text, which reinforces its circumstantial realism through frequent repetition, the picture of Crusoe repeats information supplied in writing, and functions as a visual equivalent of the text (Barr 1986: 14). Meticulous attention to detail, in this engraving, assures the reader that artistic licence is surrendered in the interest of faithful attention to 'fact'. In the nineteenth-century abridgement, repetition is surrendered in the interests of brevity, although the clean, factual style is maintained, preserving the sense of reality, if simplifying it to fit the shorter, juvenile format.

Crusoe's island is literally liminal, as a marginal zone in which the hero experiences a rite of passage.[39] The island inverts but does not, in the nineteenth-century context, ultimately subvert the social order of the adventurer's home country, since the adventure narrative confirms that social order. Although 'strange' and 'surprizing', as the original title puts it, the island adventures of Robinson Crusoe map a conservative vision of Britain and empire, identity and geography.

Mapping Crusoe

Robinson Crusoe could be read, in the nineteenth century, as a metaphorical map of Britain. British identities are mapped, through the figurative imagery of the hero, in the geography of his adventure. Crusoe, symbolically reborn as he enters the island, starts out as a man with almost no identity, and is constructed as the story proceeds. Robinson Crusoe devotes a single sentence to introducing himself and his family (quoted above). He gets straight on with his story. The brevity of this introduction is partly due to a (sometimes) spare, eighteenth-century writing style, mediated by close editing and ruthless abridgement, but it is also characteristic of adventure more generally. Since the adventurer is defined primarily by his actions, it is not possible to introduce him in much detail before the action part of his story really begins (Zweig 1974). Like all quest adventurers, Crusoe's 'perilous journey' leads to 'crucial struggle' and finally to 'the exaltation of the hero' (Frye 1990: 186–188; Green 1979). Unlike the ageless classical adventurers who endured and struggled endlessly, Crusoe ages and invents himself on the island. His identity as a white, middle-class, Christian, British man is mapped and affirmed through the course of the story.

But mapping Crusoe is not a matter of inventing identities from scratch. More conservatively, it involves recasting and reasserting existing identities. Crusoe is not washed up naked, and he clings to his cold, wet clothes, the trappings and symbols of his civilisation. Later, despite the hot sun, he is 'unwilling to be quite naked' (Anon. undated a: 44). Of 'good family', the middle-class castaway also finds himself with a shipwreck full of useful capital (tools, provisions and other useful things) and an island to himself. He is still a man, rather than a 'universal' human being. And he also has a Bible (three, in fact), which he reads and learns to understand and value. Robinson Crusoe takes these elements of Britain, of his British social self, and transplants them to the island, where he amplifies them, in himself and in his engagements with the island. Thus, when Crusoe was removed from society, society was not removed from him.

The island and its native inhabitants are vehicles in the mapping of Crusoe. The adventurer's spiritual transformation illustrates this point. Crusoe's act of rebellion against his father symbolises his original sin (Pearlman 1976). He sets out 'without letting [his] father know the rash and disobedient step [he] had taken', driven only by his 'roving disposition', guided by nothing but his sinful nature (Anon. undated a: 5). What he does, first in his 'wicked' years of adventurous wanderings, and then in the twenty-eight years of mostly solitary island life, defines who he is, and what he becomes. Crusoe's survival, after the storm and shipwreck, symbolises his forgiveness and Christian rebirth. Seeing the stranded vessel so far off shore, the morning after he is washed ashore, Crusoe 'began to thank God' for his 'happy deliverance' (Anon. undated a: 24). The waves, which have consumed the other crew-members, and are soon to consume the ship itself, leave Crusoe 'half *dead* with the water [he] took in' (Anon. undated a: 22, my emphasis). This, the adventurer's twenty-sixth birthday,

31

is also the moment of his symbolic rebirth, so that following twenty-six years of 'wickedness' his solitary life begins (Anon. undated a: 6). Crusoe's life on the island, initially solitary, leads him to Christian knowledge. Opening the Bible, the first words that meet his eye are these: 'Call upon me in the time of trouble, and I will deliver thee' (Anon. undated a: 36–37). Alone, on the island, Crusoe kneels down and prays his first prayer, that God would deliver him in time of trouble. Crusoe's 'troubles' include adventures on and around the island. Some of his first adventures take place as he explores. For example, while exploring the island's circumference in a sailing boat he made himself, Crusoe sails into a current and is 'carried along with such violence' that 'I now began to give myself over for lost' (Anon. undated a: 45). Eventually, though, he manages to return to the island, where, 'When I was on shore, I fell on my knees and gave God thanks for my deliverance' (Anon. undated a: 47). Thus the island and the surrounding waters are vehicles of Crusoe's spiritual transformation. So are the native inhabitants, cannibals, who visit the island years after Crusoe's arrival. Crusoe rescues a 'poor creature' who is about to be eaten by the 'savage wretches' (Anon. undated a: 59, 52). He later converts the man, whom he calls Friday, to Christianity, and teaches him to wear clothes and to eat goat rather than human flesh. Friday's conversion is a measure of Crusoe's Christian zeal and spiritual maturity. The other cannibals are also affected by Crusoe's religious fervour. After resolving not to judge their seemingly 'savage' practices, Crusoe decides to save a white 'christian' who is about to become their next meal. Aiming his musket, and having Friday do the same – telling him 'do as you see me do' – he kills and scatters the cannibals (Anon. undated a: 68). At first only two drop, while the others 'ran about screaming and bleeding'; eventually seventeen are killed (Anon. undated a: 70). The cannibals, like Friday and the island, are seen from Crusoe's perspective as vehicles of his own spiritual growth, people and places encountered on his spiritual journey.

Mapping the island

Robinson Crusoe maps nineteenth-century colonial geography, the British Empire in particular. Crusoe's island and its native inhabitants are vehicles for the adventurer's personal growth, for his spiritual, moral and social reflections, but they also represent, map and imaginatively colonise real places and peoples, real colonial geographies. Defoe wrote *Robinson Crusoe* with a particular colonial project in mind – British colonisation in Spanish America – which is why he located the island near the mouth of the Orinoco River. The British never colonised Spanish America quite as Defoe wanted them to, although Britain's sphere of imperial influence did extend throughout much of the region in the nineteenth century (Hyam 1993), but *Robinson Crusoe* did influence the course of British colonialism in a more general way. Although originally set specifically in the Caribbean, Crusoe's island came to be regarded a more generic, colonial space (Hulme 1992). *Robinson Crusoe* was a map of British imperial geography and a myth of British imperialism.

Transforming the island, Crusoe maps a specifically nineteenth-century brand of colonialism associated with emigration and settlement. Suddenly cut off from his past and thrust into a terrifying new existence, the storm-tossed and ship-wrecked Crusoe presents an image of modern, colonial experience. The storm, an image of disorienting modernity (Landow 1982), of powerful forces that Crusoe can neither control nor understand, sweeps him from all that is familiar. He is thrown into a new world, or rather a space in which to invent a new world. In the uncomplicated, unknown and initially unpopulated island, he is cast ashore with a little capital – the contents of the wrecked ship – and given the opportunity to make a new world, a place in which to live. He begins by exploring, imaginatively mapping the island, filling its blank spaces with names. When Crusoe saves a native islander, he also names him (Friday), identifying him with nature (which the God of Genesis created before Adam and Eve, on Saturday) and imaginatively colonising him too.[40] To Crusoe, names bring the island and its native inhabitant into existence, simultaneously colonising them and sketching 'the shadowy outline of a place'.[41] The island and Friday become terms in Crusoe's world view, settings and characters in a colonial encounter, defined from the perspective of the colonist (Brydon and Tiffin 1993). Crusoe also physically colonises the island, building a shelter for himself, building an enclosure for goats, clearing and farming land. Crusoe, a practical colonist, makes everything he needs – and a few things he doesn't, the folly of which is quickly apparent. A cedar boat, for example, is described as a 'preposterous enterprise' (Anon. undated a: 43) from the start, and when complete it is too large to launch. For the most part, Defoe concentrates his practical energies on the sober enterprise of making a place in which to live.

Crusoe, with his initial capital and his own island, is a capitalist, his colonial adventure thoroughly *petit bourgeois*, and acceptable within a broadly *petit bourgeois* value system.[42] Robinson Crusoe was not always a colonist; earlier in life he was a merchant adventurer and a plantation capitalist, and he was shipwrecked while on a slave-trading mission (to procure slaves for his Brazilian plantation). But the adventurer's days as a trader, an investor and a slave trader are relegated to his 'wicked' youth (Anon. undated a: 6), while only his later life, as a colonist, is endorsed.[43] Abridgements of *Robinson Crusoe*, focusing on the island adventure, give little mention to Crusoe's imperial career, prior to his arrival on the island. *Robinson Crusoe*, particularly in abridged form, is not a myth of all imperialisms, but more specifically of practical *petit bourgeois* colonialism (Nerlich 1987). In the words of James Joyce, who as an Irishman saw British colonialism from the perspective of the colonised,

> The true symbol of the British [colonial] conquest is Robinson Crusoe, who, cast away on a desert island, in his pocket a knife and a pipe, becomes an architect, a carpenter, a knife grinder, an astronomer, a baker, a shipwright, a potter, a saddler, a farmer, a tailor, an umbrella-maker, and a clergyman. He is the true prototype of the British colonist, as Friday (the trusty savage

33

who arrives on an unlucky day) is the symbol of the subject races. The whole Anglo-Saxon spirit is in Crusoe: the manly independence; the unconscious cruelty; the persistence; the slow yet efficient intelligence; the sexual apathy; the practical, well-balanced religiousness; the calculating taciturnity.

(Joyce 1909, translated 1964: 25)

In the early eighteenth century, *Robinson Crusoe* was, as James Joyce put it, a 'prophecy of empire', but in the nineteenth century it became a myth, promoting popular colonialism, representing and legitimating the British Empire to the British people.

Crusoe's colonial adventure displays, and diachronically resolves, some of the contradictions of colonialism. Crusoe's adventure, like that of all colonists, is pervaded by a tension between a wandering inclination (which leads to new lands) and a settled condition (in which the adventurous spirit is abandoned and settlement ensues). Crusoe changes from an enthusiastic to a reluctant adventurer, from a wanderer to a settler. Transformed principally by his Christian conversion, he renounces his restless, rebellious ways and rediscovers the domestic virtues (Zweig 1974). Ultimately he fashions a world in which adventure is no longer necessary. His island adventure lasts only as long as it takes him to transform the unknown, to possess it as a known, domesticated, enclosed, home-like space. Nature, in all its forms, is domesticated. At first Crusoe tames one corner of the island, making a secure place for himself, where he finds refuge from the wilderness and possible dangers without. Gradually adventure becomes superfluous, as the island is known, the possible dangers unmasked (the footprint, for example) and the 'savages' killed or converted. Except for the adventures that arise in the course of his practical Christian colonisation, Robinson Crusoe's life is a quiet one, centred around his residence, which is designed for 'security from being surprised by any man or ravenous beast' (Anon. undated a: 29). After an exploring expedition, he reflects, 'I cannot express what a satisfaction it was now, to come into my own hut, and lie down in my hammock bed' (Anon. undated a: 40). The contradiction of Robinson Crusoe, 'the unadventurous hero' (Zweig 1974), was mirrored in British colonies, where colonialism was associated with adventure, perhaps inspired by adventure, but was also at odds with adventure, since settled colonialism was about transforming the unknown into the known, erasing and transforming the spaces of adventure. By the end of *Robinson Crusoe*, adventure had run its course.

Through its narrative of colonisation and transformation on and of an island, *Robinson Crusoe* represents, promotes and legitimates a form of colonialism. When, after twenty-eight years, Crusoe finally leaves the island (on a British merchant ship he has helped save from mutineers), he leaves behind him an idealised British colony, which he hands over to the group of mutineers, whose merciful punishment it is to be left there. Before leaving, he 'talked to the men, told them my story, and how I managed all my household business' (Anon. undated a: 79). Crusoe gives the men his story, the story of how he colonised an island. He intends the story to guide the new colonists, inspiring their practical acts and imaginatively framing their colonial encounters. Crusoe gave his story not only to the mutineers, but also to British readers. Some readers were surely inspired to go off to sea, to seek adventure in distant lands, perhaps to settle in Canada or Australia, and ultimately help to build an empire. 'To many [readers, *Robinson Crusoe*] has given the decided turn of their lives, by sending them to sea', wrote one Scottish observer in 1834 (Ballantyne 1834, cited by Shinagel 1975: 266–267). In the middle of the nineteenth century, another prominent British observer called *Robinson Crusoe*

> a book, moreover, to which, from the hardy deeds which it narrates, and the spirit of strange and romantic enterprise which it tends to awaken, England owes many of her astonishing discoveries both by sea and land, and no inconsiderable part of her naval glory.
>
> (Borrow 1851: 1.39)

But while *Robinson Crusoe* inspired many colonial acts, its influence was not limited to the minority of Britons who were directly engaged in such acts. To the majority of Britons who stayed at home, *Robinson Crusoe* was a powerful geographical fantasy but also a colonial myth, a myth that represented British colonialism to the British people, as well as to the colonised peoples (Brydon and Tiffin 1993). Crusoe's island was a Christian utopia, a middle-class utopia, a colonial utopia. The island and the adventurer represented Britain and British colonialism – including colonial land grabs and colonial violence – in the best possible light, conservatively legitimating, powerfully mapping. To the reader in nineteenth-century Britain, Crusoe's island became an image of Britain and the British Empire, not as they had been when Defoe wrote, but as they had become by the nineteenth century.

VICTORIAN ROBINSONADES

In addition to its relatively direct mapping of Britain and empire, *Robinson Crusoe* charted cultural space in which writers and readers were able to move, as they mapped and remapped particular world views and colonies. Appropriating the narrative and the geography of Robinsons, they wrote variations known as Robinsonades. The several hundred Robinsonades published before 1900 (Green 1990 counts 277 in Europe) include stories ranging from famous books such as *The Coral Island* to more obscure titles like *Canadian Crusoes* and *Yr Ynys Unyg* (1852), translated as *Welsh Family Crusoes* (Anon. 1857), which were read mainly in Canada and Wales respectively.[44] *The Coral Island* and *Canadian Crusoes*, in particular, sent echoes of *Robinson Crusoe* through a mid-nineteenth-century world in which both British adventure literature and the British Empire were approaching periods of great expansion.

The Coral Island: a British map of the Victorian world

The Coral Island mapped the British Victorian world, and it did so in uncompromisingly bold colours. Ballantyne's *The Coral Island* is a lively tale of three boys' adventures among cannibals, pirates and exotic islands in the South Pacific. What *Robinson Crusoe* seemed only to suggest to Victorian Britons, *The Coral Island* spelled out. It was more arrogantly ethnocentric, more fervently religious, more exuberantly adventurous, more optimistic and more racist than its predecessor.

Probably the most famous Robinsonade written and published in Victorian Britain, *The Coral Island* follows *Robinson Crusoe* closely. It has been called 'one of the many tellings of a powerful mythic narrative: the story of Robinson Crusoe' (Bratton 1990: vii). Like Crusoe, the boys in Ballantyne's tale are shipwrecked near an island, and many of the events that follow shadow those of the original Crusoe. Like Crusoe, the boys set out to explore their island, ostensibly to 'ascertain whether it contained any other productions which might be useful to us', and to 'see whether there might be any place more convenient and suitable for our permanent residence than that on which we were now encamped' (Ballantyne 1858a: 83–84). Like Crusoe, they climb the mountain to gain a view of the whole island. Like Crusoe, they set about building a boat, with which to explore and perhaps leave the island. Like Crusoe, Ballantyne's trio are Christians. Jack remembers the Biblical passage that inspired Robinson Crusoe. Like Crusoe, who knew God's promise to 'Call upon me in the time of trouble, and I will deliver thee' (Anon. undated a: 36–37), Jack remembers 'ONE who delivers those who call on Him in the time of trouble; who holds the winds in his fists and the waters in the hollow of his hand' (Ballantyne 1858a: 408). The list of similarities and parallels between *Robinson Crusoe* and *The Coral Island* goes on. Eventually, though, the differences become more striking.

Few adventure stories have been gaudier, more muscular, arrogant than *The Coral Island*, in which Ballantyne simplified and exaggerated the certainties of Victorian Britain, along with the middle-class, Christian, white, male, colonial values of *Robinson Crusoe*. *The Coral Island* was one of the first adventure stories both to be labelled juvenile literature and to have an exclusively juvenile cast (Bratton 1990).[45] But the boys'-story format did not so much explain as facilitate its simple, morally didactic presentation. The simplified world view that pervades *The Coral Island* is illustrated, particularly vividly, in Ralph, Jack and Peterkin's shark encounter (Plate 2.3). The illustration, by Ballantyne's own hand, and coloured with greens, blues and yellows in some editions, is a simpler, cruder version of a well-known image of colonial encounter, by John Singleton Copley (Plate 2.4). Although produced in the eighteenth century, Copley's *Watson and the Shark* (1778) was most widely admired and known in the nineteenth, when its subject material had become more familiar and acceptable to art critics, and when reproductions and engravings of the work were widely distributed (Rosenblum and Janson 1984). Painted in London, Copley's painting depicts

Plate 2.3 In Ballantyne's *The Coral Island* (1858), heroes Ralph, Jack and Peterkin encounter a shark, but show no signs of distress, and triumph easily.

Brook Watson, a Loyalist in the American Revolution, being attacked by a shark in Havana Harbour. The sharp-toothed beast from the deep is frightening, a deadly adversary, against whom the swimmer is vulnerable and passive. Viewers of the painting knew that he survived, but lost a leg in the ordeal. Nature, symbolised by the shark, is truly red-in-tooth-and-claw, not to be lightly dismissed. Often considered the starting point for French Romantic painting, and influential among American artists, *Watson and the Shark* was also echoed in popular culture, where the drama of Watson's adventure was celebrated, sensationalised and simplified (McLanathan 1968). Ballantyne's shark attack shares the drama and excitement of Copley's original work.[46] In both images, a shark emerges from the waters of an exotic, colonial and vaguely Caribbean[47] harbour to attack a patriotic British adventurer, and in both images the viewer knows the outcome of the story, in which the shark is defeated. In both images, realistic detail conveys a sense of geographical and historical truth, producing seemingly solid ground on which a social and political vision is constructed and naturalised. But in Ballantyne's picture there are fewer details, fewer people and fewer sharks; the narrative is less complex, as is the relationship between man and nature. Ballantyne's figures are boys, not men, and they show no signs of genuine distress. Unlike the naked Watson, encircled by sharks, they remain clothed and in control, irritated but invulnerable. Their victory is not only assured, it is enjoyed, as the boys confidently and lightheartedly triumph over whatever nature throws at them. The power relation between British man and colonial space is reversed in the transition from Copley to Ballantyne.

Comparison between *Robinson Crusoe* and *The Coral Island* underlines the lighthearted confidence of the latter. Ballantyne's three heroes, Ralph, Jack and Peterkin, embarked upon an adventure that was, above all, fun. Ralph Rover presented the book 'specially to boys' in the hope that they would 'derive . . . unbounded amusement from its pages' and 'enter with kindly sympathy into the regions of fun' (Ballantyne 1858a: iii). Unlike Robinson Crusoe, who was more a survivor than an adventurer, and who was always ambivalent towards adventure, Ralph introduces his adventure without reservations of any kind. He does not doubt that the adventure is morally sound. Peterkin looks forward to his adventure excitedly. Finding himself on the island, he declares that being a castaway is 'capital, – first rate, . . . and the most splendid prospect that ever lay before three jolly young tars' (Ballantyne 1858a: 27). Ralph, Jack and Peterkin are more enthusiastic about adventure than they are about work. Unlike Crusoe, they do not attempt to reconstruct an idealised, civilised British world for themselves, and they are initially content to live on what seems to them like a South Sea Island paradise. Industriously, they set about building a boat, but soon get bored.

> 'Come, Jack', cried Peterkin . . . 'let's be jolly to-day, and do something vigorous. I'm quite tired of hammering and bammering, hewing and screwing, cutting and butting, at that little boat of ours, that seems as hard

Plate 2.4 Watson and the Shark (1778) by John Singleton Copley. The sharp-toothed beast from the deep, an image of nature and colonial space, presents the British adventurer with a serious adversary, a deadly threat.

to build as Noah's ark; let us go on an excursion to the mountain-top, or have a hunt after the wild ducks, or make a dash at the pigs. I'm quite flat – flat as bad ginger-beer – flat as a pancake; in fact, I want something to rouse me, to toss me up, as it were. Eh! what do you say to it?'

(Ballantyne 1858a: 148)

While both *Robinson Crusoe* and *The Coral Island* describe the colonial acts of white British males, the latter is more explicitly and arrogantly colonialist. Like Crusoe, the boys gaze from the mountain top, where they see their 'kingdom lying, as it were, like a map around [them]' (Ballantyne 1858a: 65). They map the island, imaginatively possessing it, in the same way that British explorers, perched on promontories all over the world, masterfully possessed what they saw. Interpreting the 'monarch-of-all-I-survey' image, a trope of British, Victorian discovery rhetoric, professor of comparative literature Mary Louise Pratt identifies a general 'relation of *mastery* predicated between the seer and the seen' (Pratt 1992: 204). This relationship was often subtle, particularly among those travel writers (read by Pratt) who seemed to look but not touch, but this was not the case in *The Coral Island*, where seeing was explicitly equivalent with possessing the island. Compared with Crusoe, Ralph and his friends are much more sure of themselves and of their place in the world, as white British males on a colonial mission. Peterkin is cheerfully racist, confidently superior. He tells his two friends

We've got an island all to ourselves. We'll take possession in the name of the king; we'll go and enter the service of its black inhabitants. Of course we'll rise, naturally, to the top of affairs. White men always do in savage countries.

(Ballantyne 1858a: 27–28)

Predictably, the 'savages' turn out to be cannibals, whom the boys help convert to Christianity and to sedentary, more vegetarian ways. The natives, eaters of human flesh and worshippers of idols, are caricatures of savagery, emphasising those aspects of native life that most disgusted Victorian Britons (Street 1975). When finally they leave the isle of Mango, the boys leave behind them a smouldering pile of idols, which the newly converted natives have burned, and they look back with satisfaction to a place where 'natives had commenced building a large and commodious church, under the superintendence of the missionary', and where 'several rows of new cottages were marked out; so that the place bid fair to become, in a few months, as prosperous and beautiful as the Christian village at the other end of the island' (where the natives were already converted to Christianity) (Ballantyne 1858a: 437). The Christian island the boys leave behind them will be more accommodating to European imperialism. As a white trader explains to Ralph, 'the only place among the southern islands where a ship can put in and get what she wants in comfort, is where the gospel has been sent to' (Ballantyne 1858a: 277).[48] Thus British colonialism and

Christian mission, seen through the eyes of Ballantyne's roving boys, reached new levels of simplicity and arrogance.

Canadian Crusoes: Mapping a British colony

While Ballantyne's Robinsonade maps a way of seeing and a series of colonial identities, Catharine Parr Traill's maps a particular colony. And while *The Coral Island* emphasises the adventurous side of the original Crusoe, *Canadian Crusoes* develops the themes of survival and settlement. Traill's story presents and promotes British emigration and settlement in the Rice Lake Plains of what is now Ontario, Canada. Traill, a member of the literary Strickland family and sister of Susanna Moodie, the well-known author of settlement narratives such as *Roughing it in the Bush* (1852), was a British settler in Canada, as well as an accomplished field naturalist and writer. Her botanical publications include *Canadian Wild Flowers* (1868, 1895), which represented thirty-four years' work, and earned Traill a reputation as 'an astute, descriptive botanist' (Eaton 1969: 172). The author of what is 'generally considered to be the first Canadian novel for children' (Egoff 1992: catalogue entry #094), a book once 'adored by children' (Eaton 1969: 140) – *Canadian Crusoes* – Traill might also be called the mother of Canadian children's literature.

Canadian Crusoes follows *Robinson Crusoe* closely, as its title suggests it will. In the foreword Agnes Strickland (Traill's sister) draws an explicit parallel between *Robinson Crusoe* and *Canadian Crusoes*.

> Where is the man, woman or child who has not sympathised with the poor seaman before the mast, Alexander Selkirk, typified by the genius of Defoe as his inimitable Crusoe, whose name (although one by no means uncommon in middle life in the east of England) has become synonymous for all who build and plant in the wilderness, 'cut off from humanity's reach?'
>
> (Strickland 1852: vi)

In *Canadian Crusoes*, as in *Robinson Crusoe*, an initial moment of disorientation plunges the adventurer(s) into *terra incognita*, where they learn to survive, make a place to live, learn to trust God and convert a cannibal to Christianity, before finally returning home.

In *Canadian Crusoes* the generic geography of *Robinson Crusoe*'s adventure is mapped onto a particular place. Traill drew upon her colonial experiences, backed up by botanical and zoological knowledge, to give the Canadian adventure a sense of reality. Her Robinsonade is adapted to its inland setting. Whereas seafaring Crusoes, particularly in the eighteenth century, could plausibly be swept by storms onto unknown but real islands, inland crusoes in the nineteenth century could simply get lost and find themselves in unknown, island-like spaces.[49] Whereas British writers from Shakespeare to Defoe used images of storms and shipwrecks, borrowed from contemporary (and earlier) travel narratives, Traill was inspired by stories – printed in the local paper – of children lost in the wilderness (Schieder 1986).[50] As Strickland explained, 'scarcely a summer passes over the colonists in Canada, without losses of children from the families of settlers occurring in the vast forests of the backwoods, similar to that on which the narrative of the Canadian Crusoes is founded' (Strickland 1852: vi–vii).

Canadian Crusoes is the story of three children who get lost in the wilderness, and build a colony there. Hector, Louis and Catharine are distracted by wild strawberries while out in search of stray cattle. Late in the day they realise they have 'wandered from the path' and 'neglected, in their eagerness . . . to notice any particular mark by which they might regain it' (Traill 1852: 19). The children try to find their way home, following cattle trails and stream beds, but eventually they realise that their wanderings are fruitless. The are lost, disoriented, with 'no idea of distance, or the points of the compass', and no idea 'in what direction the home they had lost lay' (Traill 1852: 88–89). Weary and disoriented, they fall asleep in the forest. When they wake the next morning they are, like Robinson Crusoe on his island, in *terra incognita*. First and foremost, they must survive. They set about finding food and making shelter for themselves. They learn to survive in the particular Canadian wilderness, adapting to the plants and animals they find, which Traill renders in a detail that reflects her knowledge of Canadian natural history. Despite Traill's clear, naturalistic geographical, botanical and zoological detail, the setting is an unknown, frightening wilderness to the children.

For all its local colour and detailed natural history, the geography of Traill's adventure is a blank, unknown space, somewhat frightening but ultimately malleable, putty in the hands of white colonists who may shape it as they will. Hector reflects that 'children can do a great many things if they only resolutely set to work, and use the wits and the strength that God has given them to work with' (Traill 1852: 125). He and his companions resign themselves to their predicament, to the space in which they have found themselves. They begin to re-orient themselves: 'And now arose the question, "Where are we?"' (Traill 1852: 34). Anticipating Northrop Frye (who, a century later, was to teach Canadians to ask 'where is here?') the three young Crusoes acknowledge that they have no prior knowledge of where they are, and so must invent their world. In the face of the children's resolution – to 'do a great many things' – nature puts up little genuine resistance, and submits to their will. The geography of *Canadian Crusoes*, never the material possession of native inhabitants, is a projection of the children's geographical fantasies and colonial desires.

> Much the children marvelled what country it might be that lay in the dim, blue, hazy distance, – to them, indeed, a *terra incognita* – a land of mystery; but neither of her companions laughed when Catharine gravely suggested the probability of this unknown shore to the northward being her father's beloved [Scottish] Highlands. Let not the youthful and more

learned reader smile at the ignorance of the Canadian girl; she knew nothing of maps, and globes, and hemispheres, – her only book of study had been the Holy Scriptures, her only teacher a poor Highland soldier.

(Traill 1852: 131–132)

Nothing, not even a Canadian winter, could put up much resistance to geographical imaginations and colonial wills such as these.

In addition to suggestive images of settlement in the Rice Lake Plains area, *Canadian Crusoes* explicitly sells the region to potential colonists, reflecting the commitment to colonialism that Traill displayed in her life as a colonist, and in her other works, such as *The Female Emigrant's Guide* (1854). The children transform the space in which they find themselves, building a Christian colony. The Rice Lake Plains area has been transformed in this general way, insists Traill, by practical Christian settlers like the three young Crusoes. Her story of untamed wilderness and cannibalism is set in the past, not the present.

> At the time my little history commences, this now highly cultivated spot was an unbroken wilderness, – all but two small farms, where dwelt the only occupiers of the soil, – which owned no other possessors than the wandering hunting tribes of wild Indians, to whom the right of the hunting grounds north of Rice Lake appertained, according to the forest laws.
>
> To those who travel over beaten roads, among cultivated fields and flowery orchards, and see cleared farms and herds of cattle and flocks of sheep, the change would be a striking one.
>
> (Traill 1852: 2)

The transformation of (what is now) Ontario in Traill's narrative corresponds roughly to material change in the region, which white settlers had barely begun to colonise when Traill first arrived there in 1832. But Traill did not seek to describe, so much as to promote colonisation. In *Canadian Crusoes* she missed no opportunity to promote Ontario as a destination for emigrants, advising British emigrants of its attractions and advantages, and implicitly comparing it with rival destinations such as Australia. Strickland's foreword explicitly directs emigrants to Canada, 'our nearest, our soundest colony, unstained with the corruption of convict population; where families of gentle blood need fear no real disgrace in their alliance' (Strickland 1852: xi). Thus, Traill appropriated a conventional narrative form, of established popularity, as a vehicle for colonisation in a specific colonial setting.

EXPLORING THE SPACES OF ADVENTURE

Like the adventure storyteller, who begins with a hazy map, I have sketched the outlines of adventure's geography, space to be explored in more detail in the narrative that follows. Reading Robinsons and Robinsonades, I have raised more questions than I have resolved. I have suggested that Robinsons, Robinsonades and other adventure stories appealed to geographical imaginations and created cultural spaces in which constructions of geography and identity were mapped. But what difference does the nature of this cultural space make, specifically to identities and geographies? Is the spatial metaphor – *mapping* identities and geographies – really necessary? Following on from this, I raise the question of how maps become meaningful. The shifting meaning of *Robinson Crusoe*, highlighted by the contrast between readings in eighteenth- and nineteenth-century Britain, draws attention to the potential roles and powers of situated readers, rather than just intentional authors and texts themselves. Power relations, embedded within the situated mapping process, connect texts with identities and geographies. The shifting and contradictory meanings of *Robinson Crusoe*, the story that is radical in one context and conservative in another, with a hero who is adventurous in one context and unadventurous in another, illustrates the ambivalence of adventure and calls for a textured interpretation of its geography.

3

MAPPING MEN

Spaces of adventure and constructions of masculinity in *The Young Fur Traders*

Identities are mapped in real and imaginary, material and metaphorical spaces. Moving from *Robinson Crusoe*, a broad map of nineteenth-century Britain and empire, towards a more focused exploration of mapping, I turn to explore the mapping of identity in adventure literature. Maps naturalise and normalise, fixing constructions of identity and geography. And since they are often associated with conservative reinscriptions of the *status quo*, with enclosing and circumscribing geographies and identities, maps tend to fix dominant ways of seeing geography and hegemonic constructions of identity. Hegemonic constructions of race, gender, class and other dimensions of identity reflect the characteristics of the spaces in which they are mapped. In other words, the nature of the spaces – real and imaginative, material and metaphorical – make a difference to the nature of the hegemonic identities.

Adventure narratives, ancient and modern, describe (what Sauer called) '*man's* contest with nature', *his* encounters with the 'unknown'.[51] The same can be said of other geographical narratives; geographers such as Gillian Rose, Peter Jackson and Felix Driver have said it. Rose (1993: 4) argues that academic 'geography is masculinist', while Jackson and Driver, reading a more broadly defined geographical literature ranging from advertising copy to tourist guides and travel books, come to similar conclusions (Driver 1992; Jackson 1991). Sauer's gendered language is appropriate, since for the most part the eclectic geographical literature of adventure is restricted to *male* encounters – real and imaginative – with nature and the unknown. Thus Martin Green calls adventure a 'masculinist' literature, (ostensibly) for, about and by men and boys (Green 1979). Adventures map masculinities in relation to geography, and geography in relation to masculinities, as I demonstrate in this and the next chapter.

Adventures map masculinities, not masculinity. Since masculinities are heterogeneous, and since adventure stories are told and written differently in different times and places, the masculinism of adventure is historically and geographically contextual, rather than general. Adventure stories chart masculinities contextually, in relation to particular constructions of class, race, sexuality and other forms of identity and geography.[52] The masculinities mapped in British Victorian

45

adventure, then, reflect a range of British Victorian identities and geographies. They also reflect the specific form taken by adventure in that context.

British Victorian adventures, although reminiscent of myths and legends that were told previously, perhaps 'always' (Campbell 1949), are historically distinctive, departing from tales of old. In the Victorian period, for the first time, adventures were printed in large quantities and read by mass audiences – not only in Britain, but around the world, especially the British Empire. Also for the first time, adventure stories were commonly classified as juvenile fiction. The precedent for British juvenile adventure fiction, first produced in the middle of the nineteenth century by writers such as Robert Ballantyne, Captain Marryat and Mayne Reid, was earlier adult (or not age-specific) adventure, which had become popular with young readers, many of whom were bored with contemporary children's literature. Modern adventure classics by Daniel Defoe, Jonathan Swift and James Fenimore Cooper, none of them originally intended specifically for juvenile audiences, were re-issued (sometimes abridged, edited, illustrated) and re-marketed as children's literature. In some cases, sentences were shortened, vocabulary simplified and didacticism emphasised, adapting books for the juvenile market (although not all children's books are short, simple or didactic), while in others books were simply reclassified. Adventure fiction, not originally or intrinsically juvenile, was generally redefined as juvenile literature.

Within the context of British Victorian juvenile adventure literature, I focus on the Scottish writer Robert Michael Ballantyne (1825–1894), whom I introduced in Chapter 2, paying particular attention to his novel *The Young Fur Traders* (1856, originally entitled *Snowflakes and Sunbeams; or, The Young Fur Traders, a Tale of the Far North*) (Plate 3.1).[53] I select Ballantyne because of his influence in contemporary Canadian adventure fiction and his fame in Britain. I examine a single novel in detail, situating it within its genre rather than always generalising about that genre, because adventure stories were so conventional, their plots and characters so formulaic and familiar, that a single novel can be called representative. Ballantyne was popular with Victorian children, particularly boys. Boy fans would often follow the author around, badgering him for attention and autographs.[54] A survey conducted among British schoolboys in 1884 placed him well inside the ten most popular writers, alongside other adventure writers including W.H.G. Kingston, Jules Verne and Captain Marryat (Salmon 1886a, 1888) (Table 3.1).[55]

Today Ballantyne is best remembered as the author of the island adventure story *The Coral Island* (1858a), but in his lifetime he was associated – as an author, an illustrator and a 'real life adventurer' – less with the South Pacific than with British North America (Plate 3.2). He began his literary career with a memoir of the five years he spent as an apprentice with the Hudson's Bay Company (HBC), entitled *Hudson's Bay; or, Every-Day Life in the Wilds of North America* (1848). Although he returned from the HBC Territories to Scotland via New York, where he acquired the clothes and haircut of an urban dandy, Ballantyne later cultivated a manly, Canadian look. The bearded adventurer came to embody a theatrical

Plate 3.1 Title page of *Snowflakes and Sunbeams; or, The Young Fur Traders, a Tale of the Far North*, originally published by Nelson in 1856, based on artwork by the author. Snow-shoes, rifles, sleds, canoes and entrapped animals are all icons of Canadian fur-trading life, and of boyish pleasure and adventure.

Table 3.1 Favourite authors

Author	Boys (790)		Girls (1210)	
	no.	%	no.	%
Charles Dickens	223	28	335	28
W.H.G.Kingston	179	23	19	2
Walter Scott	128	16	248	20
Jules Verne	114	14	22	2
Captain Marryat	102	13	5	0.5
R.M. Ballantyne	67	8	6	0.5
Harrison Ainsworth	61	8	0	0
William Shakespeare	44	6	75	6
Mayne Reid	33	4	0	0
Lord Lytton	32	4	46	4
Charles Kingsley	28	4	103	9
Daniel Defoe	24	3	8	1
J. Grant	12	2	0	0
J.F. Cooper	12	2	0	0
C.M. Yonge	0	0	100	8
Mrs Henry Wood	0	0	58	5
E. Wetherell	0	0	56	5
George Eliot	9	1	50	4

Source: Salmon (1888)
Note: Respondents were asked, 'Who is your favourite author?' and 'Who is your favourite writer of fiction?' Survey conducted by Charles Welsh in 1884, analysis and report by Salmon, published in 1888. Salmon amalgamated responses to both questions, so total number of responses are greater than total number of respondents. Salmon's report does not include demographic data (age, social class of individual or school catchment area, regional breakdowns), which might have indicated representativeness and accuracy. He states that, of 2000 respondents, 790 were boys; this means there were 1210 girls, although Salmon does not say so explicitly.

Canadian manliness, which appealed to the crowds who paid not only to read his books, but also to see him on stage, singing *voyageur* songs and telling tales of life in the wilds of British North America (Quayle 1967). His first novel, *The Young Fur Traders*, was published in 1856, and went through multiple reprintings and editions, remaining in print for about a century, with occasional printings thereafter.[56] Another popular book by Ballantyne that appeared that year, *The Three Little Kittens* (1856), was published under a pseudonym, which protected the author's manly image (Comus 1856, 1857). *The Young Fur Traders* established Ballantyne's style, which was maintained with little variation in his other adventure stories, particularly the thirteen he set in North America. For example, the cast of characters in *The Young Fur Traders* – including a teenage boy, a 'manly' man, a white trapper, a 'good' and a 'bad' Indian – is replicated almost exactly in other Ballantyne tales such as *The Pioneers* (1872) and *The Prairie Chief* (1886). In *The Pioneers; a Tale of the Western Wilderness*, the *dramatis personae* include a 16-year-old male hero and his manly father, as well as a conventionally 'good' Indian and his wife. And in *The Prairie Chief* (1886) there is a white trapper, a converted Indian and his wife, and a Wesleyan missionary.[57]

48

Plate 3.2 Robert Ballantyne in Scotland (source: Quayle 1967). Ballantyne returned from Canada via New York, where he acquired clothes and a haircut that would have identified him, if anything, as a young, urban dandy. His rustic, manly, Canadian look was cultivated later on, in Scotland.

In order to produce at least two full-length tales a year – over seventy during the course of his writing career – in addition to his other short stories and works of juvenile non-fiction, Ballantyne had simply to apply the formula he established in *The Young Fur Traders*. The formula also worked for other writers of Canadian adventure stories in the Victorian and Edwardian periods. According to Sheila Egoff, the Canadian juvenile literature critic, 'most later writers of the [Canadian] boy's adventure story strove for the Ballantyne style' (Egoff 1992: catalogue entry #112).[58] And the basic formula worked in adventure stories, by Ballantyne and others, set outside North America. A detailed, contextual reading of *The Young Fur Traders* should therefore inform histories of adventure and masculinist geography more generally.

I begin by considering the social context of Victorian Britain, in which boys' adventure stories and other gendered juvenile fictions were produced and consumed, and then focus upon the ways in which the setting of *The Young Fur Traders* enabled Ballantyne to articulate an ideology of British masculinity. I interpret the hero's journey through the spaces of adventure as a journey through adolescence, a rite of passage from boyhood to manhood, and show how his masculinity is mapped in relation to the spaces of his adventure.

MASCULINITY IN VICTORIAN BRITAIN: JUVENILE LITERATURE

Ballantyne asserted his ideology of masculinity in the context of Victorian Britain. To understand the ways in which he mapped masculinity, it is first necessary to know something about the popular cultural context in which his story was produced and consumed. In this section I contextualise *The Young Fur Traders*.

In the Victorian period, for the first time in Britain, magazines and books (other than the Bible and the occasional pictorial chapbook) were purchased and read by mass audiences, which included children. Publishers responded to the emergence of a largely new social, demographic and literary phenomenon: mass juvenile literacy (Schofield 1981; Vincent 1989). The growth of literacy among young people was the product of many forces, including high birth rates, general material improvement, and wider availability of primary education, initially at Sunday school (Bratton 1981; Dunae 1989). As the British population grew, the nation developed a youthful population profile. The number of under-14s nearly doubled between 1841 and the turn of the century, rising from 5.7 to 10.5 million, and never falling below a third of the total population. Many of these young people – a higher proportion than ever before – went to school and learned to read. The trend towards mass juvenile literacy was long established by 1870, when an Education Act ensured universal access to elementary education; ten years later another Act made such education compulsory. Successive Acts increased the number of years a child spent in education, raising the basic school-leaving age from 10 to 11 years of age in 1893, and to 12 in 1899 (Walvin 1982).

Britain's state schools not only provided a secular education, they also delayed entry to the work force and helped delay adulthood in general. These factors contributed to the rapid expansion of a juvenile literature market. Whereas early nineteenth-century children, deprived of entertaining children's literature, tended to read adult literature (if they read at all), their counterparts in the latter part of the century enjoyed a literature of their own. In 1888 one observer was able to remark that 'Whoever undertakes to write the literary history of England during the latter half of the Nineteenth Century will be confronted by a force hitherto almost non-existent. The floods of books for boys and girls with which the approach of Christmas has in recent years been heralded, were unknown four decades ago' (Salmon 1888: 32) (Plate 3.3).

As its market expanded, children's literature became competitive and specialised, with increasingly separate, increasingly gendered boys' and girls' literature (Dunae 1989). Whereas the characters of early nineteenth-century juvenile literature were children, their successors in the second half of the century were boys and girls (Egoff and Saltman 1990). Characters such as Goody Two-Shoes[59] and Jemmy Studious, early-Victorian remnants of the essentially seventeenth-century Puritan children's storytelling tradition, were generally ascetic, angelic and (according to contemporary conceptions of femininity and masculinity) more feminine (pure, asexual) than masculine (sexual, aggressive) (Nelson 1991). Later nineteenth-century characters, in contrast, were polarised along gender lines, becoming clearly masculine or feminine. The masculinisation of boys was partly in response to changing ideas about what it meant to be feminine. Claudia Nelson (1991) interprets the masculinisation of boyhood in the light of contemporary developments in psychology. As psychologists began to think of women as actively sexual, in the second half of the century, it became impossible to think of feminine boys as asexual, hence sexually pure, and (later on) possible to think of them as (in some way) homosexual – which writers were hardly lining up to do (Nelson 1991). Nowhere was the polarisation of gender more pronounced than in adventure fiction, with its lively, overtly masculine heroes such as Frank Fearless and Dick Dare, and its general absence of heroines, girls' stories and women writers.

So-called boys' stories were explicitly addressed and marketed to boys, although they were also read by men, women and girls. A typical preface (Mayne Reid's preface to *The Boy Hunters*) stated that 'For the boy readers of England and America this book has been written, and to them it is dedicated' (Reid 1853: v). Critics and reviewers judged adventure books with boy readers in mind. *The Athenaeum*, for example, expected *The Young Fur Traders* to be 'the delight of high-spirited boys' (Quayle 1967: 106). Boys' stories were, in general, most popular among boys. For example, *Robinson Crusoe* and *Swiss Family Robinson* were the most popular books among boys surveyed in 1884, and were the favourites of far fewer of the girls who responded to the same survey.[60] Nevertheless, contemporary critic Edward Salmon (1888: 28) suggested that 'if we were to take the country through, we should find that nearly as many girls

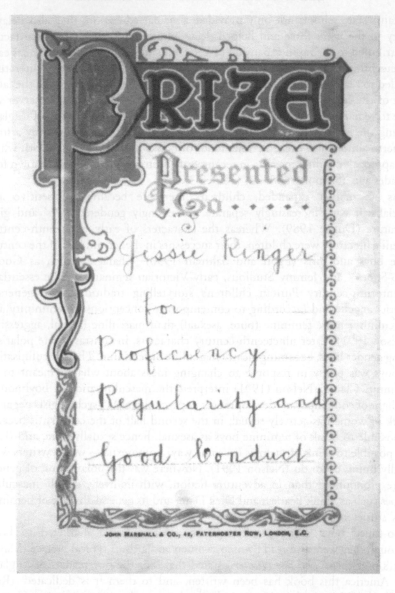

PRIZE

Presented
To......

Jessie Ringer

for

Proficiency

Regularity and

Good Conduct

JOHN MARSHALL & CO., 42, PATERNOSTER ROW, LONDON, E.C.

Plate 3.3 Books as prizes and presents. Inscriptions show that children often received rather than chose or bought many of the books they read. Parents, other adult relatives, Sunday and secular day school teachers all selected books for children, with an eye on their moral as well as entertainment value. Jessie Ringer, for example, received a 1906 edition of Ballantyne's *The Dog Crusoe*, first published by Nelson in 1861, as a Sunday school prize.

as boys have read *Robinson Crusoe, Tom Brown's School Days*, and other long-lived "boys' stories'". *The Boy's Own Paper*, the most popular magazine among boys, was also the second most popular magazine among girls (Salmon 1888). One girl explained the attraction of boys' stories, complaining that

> People try to make boys' books as exciting and amusing as possible, while we girls, who are much quicker and more imaginative, are very often supposed to read milk-and-watery sorts of stories that we could generally write better ourselves.

(Salmon 1888: 29)

Victorian girls and women found traces of adventure in school stories and other popular literary genres not defined principally as adventure, but those who wanted to read lively adventures usually turned to so-called boys' stories (Jenkinson 1946). In magazines like *The Boy's Own Paper* they found excitements – 'bloodthirsty things and adventure and boys', as one female reader put it[61] – that were not available in *The Girl's Own Paper*, or other girls' literature. Girls read boys' literature partly because tales of girl and women adventurers, which had been relatively common in the past, became rare. Girls' adventures virtually disappeared from print by the middle of the century (Wheelwright 1987), while tales of girl adventurers lagged well behind those of boys in the new juvenile literature. It was not until early in the twentieth century that girls' adventure stories – by writers such as Bessie Marchant – began to compete effectively for the attentions and the pocket money of British girl (and some boy, men and women) readers (Cadogan and Craig 1986; Major 1991; Reynolds 1990; Turner 1957: 12). But throughout most of the Victorian period, so-called boys' adventure stories and the masculinities they constructed were not just about boys and men, but girls and women too.

As its market expanded, children's – especially boys' – literature became less piously ascetic and more exciting. Christian publishers, including Evangelical organisations such as the Religious Tract Society (RTS) and the Society for the Propagation of Christian Knowledge (SPCK), which were traditionally favoured (and distributed) by Sunday schools, and generally faithful to long-standing traditions of Puritan storytelling, faced secular competition (Allen and McClure 1898; Lowther Clarke 1959).[62] As early as 1824, the RTS expressed dissatisfaction with its market share, and determined to win back readers, some of whom were distracted by more entertaining publications available elsewhere. Commercial publishers had made their 'penny dreadfuls' (low-brow, often violent magazines) less dreadful in order to appeal to the parents and teachers who, they knew, exercised some control over their children's reading. Edwin J. Brett, founder of the Newsagents' Publishing Company, which had specialised in the fiercest of 'bloods', launched the comparatively tame *Boys of England* in 1866, announcing that his aim was to 'enthral . . . by wild and wonderful but healthy fiction' (Turner 1957: 67). Christian publishers began to recognise that they would not control the juvenile market with traditionally austere, moral tales alone; that if they did not make their stories more exciting and attractive (according to contemporary tastes) they would lose readers, influence and customers (Bratton 1981). The SPCK, RTS and other Christian publishers such as Thomas Nelson (Ballantyne's main publisher) began to produce attractive, illustrated books and magazines. Samuel Orchart Beeton launched Britain's first Christian response to the changing magazine market. *The Boy's Own Magazine* (1855) achieved a circulation in the tens of thousands, modestly successful by contemporary standards, but dwarfed by the success of *The Boy's Own Paper*, which was launched by the RTS in 1879. At its peak, *The Boy's Own Paper* (known to many as *BOP*) achieved a circulation above one million (Turner 1957: 66; Dunae 1976; Cox 1982). Five years after its first publication, *BOP* was the most popular magazine among boys, and a favourite among girls (Table 3.2). The house style of *BOP* was epitomised in the lively, Christian stories of Ballantyne, one of the magazine's regular contributors, whom the *BOP* once praised for being 'at once amongst the manliest of men and the sincerest of Christians' (*The Boy's Own Paper* 1893–4: 429).

The combined forces of gender polarisation and commercialisation helped to produce 'new boys' – in the form of heroes and, perhaps, readers. These new boys made their debut in the mid-century school and adventure stories of Christian Socialists. Christian Socialists – including J.M. Ludlow, F.D. Maurice, Charles Kingsley and Tom Hughes – used boys' stories as a medium in which to voice their reaction to contemporary social problems in Britain (Vance 1985). Departing from the asceticism of traditional Puritan juvenile literature, and from the alleged asceticism of contemporary Roman Catholic and Evangelical Christianity, they asserted a physical, worldly 'Christian manliness' (also known as 'muscular Christianity'). They did not intend Christian 'manliness' – short for humanliness – to be an exclusively male idea (Haley 1978); they were concerned not just with male bodies but with (the physical well-being of) all bodies. In practice, however, they spoke about men (not humans), about manliness (not humanliness) and about male bodies (not all bodies). The Working Men's College (London), of which Maurice was the principal and Hughes the boxing instructor, was known for its 'austerely masculine atmosphere', as was the Christian Socialist imagery of health: the imagery of male athleticism (Masterman 1963: 169). Christian Socialists spoke about athletic males and they idealised athletic, 'manly' sports, which were becoming popular (or, at least, widespread) in British schools – and school stories. In Victorian juvenile litera-ture, where Christian instruction of one form or another was standard, it was the physical rather than the spiritual dimension of muscular Christianity that was most striking to readers. In *Tom Brown's School Days* (1857), and specifically in the figure of Tom Brown, Hughes invented the muscular Christian hero (in various guises) of over half a century of boys' school stories.

Muscular Christians were transplanted from school to adventure stories, although they were not transplanted intact. They were changed by the settings of adventure. Christian manliness, comparatively androgynous and heterogeneous in school stories such as *Tom Brown's School Days*, was caricatured and simplified

Table 3.2 Favourite magazines

Magazine	Boys (790)		Girls (1210)	
	no.	%	no.	%
The Boy's Own Paper	404	51	88	7
Tit Bits	27	3	0	0
The Standard	20	3	0	0
The Union Jack	16	2	4	0.5
The Boy's World	16	2	0	0
Girl's Own Paper	0	0	315	26
Little Folks	7	1	71	6
Cassell's Family Magazine	0	0	35	3
Quiver	0	0	29	2
Other/Not stated	300	38	668	55

Source: Salmon (1888)
Note: Respondents were asked, 'Which is your favourite magazine?' Responses may be biased by the environment in which this survey was conducted: children's classrooms. In the presence of their teachers, children may have been less likely to mention their less respectable favourites. See note for Table 3.1.

in adventure stories. Nelson (1991) argues that Tom Brown's physical masculinity was countered by the angelic 'femininity' of his friend George Arthur, and by a mixed cast of other male characters. Adventure heroes, in contrast, were uncompromisingly, uniformly masculine and, as I shall argue with reference to *The Young Fur Traders*, their new masculinity was made possible and plausible by the settings in which their adventures took place. Writers such as Ballantyne did find British settings for adventure stories, including the tough, all-male settings of the railways and tin mines, and among firemen and lifeboat crews. But while one *might* have adventures and become (what Ballantyne regarded as) manly in Britain, one was more likely to do so in more exotic spaces of adventure, overseas. As one adventure hero explained, in the Canadian adventure/travel book, *By Track and Trail*[63] (1891):

> I am determined not to go back to England, to be a drudge in an office, in a bank, or something of that sort, the very thought of which disgusts me. Just think of what most of those fellows are at home; they spend one half their lives at a desk, the other half fadding about their dress or their appearance. Why, they are mostly as soft as girls, and know nothing but about dancing, and theatres, and music-hall singers.
>
> (Roper 1891: 116)

Although he concedes that many men who stayed in England were 'just as manly as you can desire', Tom insists that 'there are far too many, and the better-off ones, too, who just disgust me, and I will not go back and run the slightest chance of becoming one of them' (Roper 1891: 116). British masculinity, not transplanted, is constituted in the geography of adventure.

55

A GEOGRAPHY OF MASCULINITY:
RITES OF PASSAGE IN *THE YOUNG FUR TRADERS*

Journeys of male heroes, through the spaces of *The Young Fur Traders*, are metaphorical journeys from boyhood to manhood – white, middle-class masculine rites of passage. The boy heroes of Victorian boys' adventure stories, unlike their counterparts in most earlier adventure narratives (right up to *Robinson Crusoe, Swiss Family Robinson* and *Masterman Ready*, all with adult heroes), do not start out pre-formed, timeless heroes (Bakhtin 1981). They do not set out with manly qualities, but acquire them *en route*. The adventurer's manhood is constructed, naturalised and normalised in and through the setting, as I now argue with reference to *The Young Fur Traders. The Young Fur Traders* describes the adventures of 15-year-old Charley Kennedy, and his 14-year-old friend, Harry Somerville. Charley and Harry encounter other *dramatis personae* in the story, including a *voyageur* named Jacques Caradoc; Hamilton, another young HBC apprentice; a Christian Indian known as Redfeather; his adversary, Misconna; and wife, Wabisca. In *The Young Fur Traders*, Ballantyne constructs a form of Christian manliness, which bears clear traces of the story's setting.

The setting of *The Young Fur Traders* is 'far removed from the abodes of civilized men' (Ballantyne 1856: 10). It is liminal space in which 'civilized' men's rites of passage take place. Charley, hero of the story, sets out in 'that ambiguous condition that precedes early manhood' (Ballantyne 1856: 9) – he is adolescent. Adolescence – a new concept and social reality in Victorian Britain (and North America) – was initially the privilege of middle-class boys, since it required the luxuries of extended education or deferred employment.[64] Charley's middle-classness is signified by his privilege to choose between a business career and a life of adventure, and by his well-spoken accent, which contrasts with that of the stable boy, for example. What Charley calls a 'new horse', the stable boy calls a 'noo 'oss', and what Charley rides, the stable boy mucks out (Ballantyne 1856: 31). Adolescence is not available to the working-class boy, nor to the girl in the story, Charley's sister. So while Charley is leaping around the HBC Territories, learning the ways of men, his 14-year-old sister Kate is keeping house and cutting tobacco for her parents. The privilege of adolescence gives the middle-class boy access to the geography of adolescence, the liminal space in which his rite of passage may take place. Charley's journey takes him to a land where the rules of his 'civilized' society (in Britain and the settled regions of British North America) do not apply. The story 'plunges' him 'into the middle of an Arctic Winter; conveys him into the heart of the Wildernesses of North America' (Ballantyne 1856: 9). Here, he is able to behave in ways that would not be possible at home. He is able to live 'rough', developing his 'body and spirit' in ways that 'an easy life' would not permit. He explains, in a letter to Harry (Ballantyne 1856: 200–201), that:

> 'Roughing it' I certainly have been, inasmuch as I have been living on rough fare, associating with rough men, and sleeping on rough beds under the starry sky; but I assure you, that all this is not half so rough upon the

constitution as what they call leading an *easy life*, which is simply a life that makes a poor fellow stagnate, body and spirit. . . . I am thriving on it; growing like a young walrus; eating like a Canadian *voyageur*, and sleeping like a top.

After a time spent wandering in the wilderness, learning from the 'rough life', Charley can expect to return – literally and metaphorically – to the 'abodes of civilized men', as a privileged member of that society.

The setting of *The Young Fur Traders* is *terra incognita, carte blanche* in which it is possible freely to invent and reinvent adventures and adventurers – including the masculinity of adventurers. It is a vast, unnamed region in the colonial geographical imagination, both inside and outside this particular adventure story. The geography of the HBC Territories was largely 'unknown' to many Britons and non-native British North Americans until the mid-1850s, when high-profile exploring expeditions went there. Previously, explorers and fur-traders charted the broad outlines of the region, although the HBC, which provided map-makers with much of their data, was notoriously cagey about what it knew.[65] Allegedly, it preferred to keep valuable geographical knowledge to itself, allowing the Territories to languish in its possession and remain a hazy region in the popular geographical imagination (Owram 1980). Ballantyne emphasised this haziness. From Fort Garry (the HBC fort situated where Winnipeg stands today), 'the great prairie rolled out like a green sea to the horizon, and far beyond that again to the base of the Rocky Mountains' (Ballantyne 1856: 17). By pointing to a vast region but not naming it, Ballantyne defines a sublime, silent space (see Eagleton 1990). By refusing to name, to articulate the region, he preserves its openness. To 'articulate west', suggests Canadian Studies Professor William New, would be to undermine its sublimity, since 'to find words to articulate it, is paradoxically at once to create and limit it. In the act of articulation, the endlessness of possibility is circumscribed, for an actual identity is announced' (New 1972: xii). Against a backdrop of silence, it is possible to spin yarns, and to invent identities.

Charley enters trackless wilderness, space rather than place, a land of directions rather than destinations, which is neither domesticated nor circumscribed, as it would be by names and detailed maps.[66] Particular geographical details are kept out of the story. Harry, excitedly joining Charley and the *voyageurs* as they prepare to head north into Lake Winnipeg, does not have any geographical details worked out. 'I'm going all the way, and a great deal farther. I'm going to hunt buffaloes in the Saskatchewan, and grizzly bears in the – the – in fact everywhere!' (Ballantyne 1856: 80). Jacques Caradoc, an older hunter and *voyageur* with whom the young *voyageurs* travel at various times, has had many years of experience in the northwest, but he remains equally uninterested in the details of its geography. He explains (Ballantyne 1856: 338):

To hunt, to toil in rain and in sunshine, in heat and in cold, at the paddle or on the snowshoe, was his vocation, and it mattered little to the bold hunter whether he plied it upon the plains of the Saskatchewan, or among the woods of Athabasca.

Ballantyne's landscapes are generic rather than particular, vaguely romantic – with 'wild primeval forests' and mountains bathed in the 'rich glow of red' light – and reminiscent more of European literary and artistic landscape clichés and conventions than of anything specifically western Canadian (that may also be clichéd and conventional) (Ballantyne 1856: 171–173). His prairies, for example, are metaphorical oceans, clichéd landscapes. In the winter, the prairie is 'a vast sheet of white . . . broken, in one or two places, by a patch or two of willows, which, rising, on the plain, appeared like little islands in a frozen sea' (Ballantyne 1856: 40). The debt to writers such as James Fenimore Cooper (and his imitators) is clear.[67] Cooper, comparing the prairie earth to an ocean, saw, for example, 'the same waving and regular surface, the same absence of foreign objects, and the same boundless extent of the view' (Cooper 1827: 12). Ballantyne occasionally acknowledged his influences, for example prefacing a description of ocean-like prairie with the acknowledgement that 'The prairies have often been compared, most justly, to the ocean' (Ballantyne 1861a: 71). But his derivative geography is not necessarily evidence that Ballantyne was unable to see or describe western Canada, nor that he was handicapped with 'eastern eyes' (to borrow a phrase from Canadian literary historian Dick Harrison 1977). On the contrary, Ballantyne's generic landscapes and vague geographies serve the purposes of his story, in defining a setting that is malleable, open to the adventurous imagination.

Ballantyne's general interest in *terra incognita*, space in which geographical fantasies were possible, is spelled out most clearly in another story. The setting of *The Pioneers*, a tale based on Alexander Mackenzie's explorations of British American wilderness, is 'almost *terra incognita*' (Ballantyne 1872: vii). Readers are invited to imagine the explorer in a remote HBC outpost at the northern end of Lake Superior. The unmapped space around him is amenable to Mackenzie's geographical fantasies, and to the geographical fantasies of the author and his readers.

> Alexander Mackenzie – while seated in the lowly hut of that solitary outpost poring over his map, trying to penetrate mentally into those mysterious and unknown lands which lay just beyond him – saw, in imagination, a great river winding its course among majestic mountains towards the shores of the ice-laden polar seas.
>
> (Ballantyne 1872: 32–33)

The space that Ballantyne and other writers could plausibly refer to as *terra incognita* diminished gradually over time, although they continued to find settings that were 'imperfectly known to the geographer', as one writer put it (Daunt 1882: 13). Here, off the map, geographical imaginations were set free.

In a setting defined by movement and freedom, the hero defines himself through his actions. Charley is almost always in motion, and his movements – on horseback, in snow-shoes, with a dog sledge, in a canoe and on foot – supply most of his story's action, as well as its minimal, linear structure. Charley's first adventure, in which he joins a wolf hunt – on a horse that is judged to be as wild as a buffalo – illustrates the sense of movement and excitement that pervades *The Young Fur Traders*. 'With his brown curls streaming straight out behind, and his eyes flashing with excitement, his teeth clenched, and his horse tearing along more like an incarnate fiend than an animal', Charley gallops across the prairie. Through this action, Ballantyne celebrates boyishness, and suggests the necessity for boys boldly and blithely to enter into exciting, dangerous adventures, which will enable them to discover their potential and reach their limits, as young men. This message, implicit in *The Young Fur Traders*, is stated explicitly in another story by Ballantyne, *The Gorilla Hunters*, where the hero reflects that:

> [Boys] ought to practise leaping off heights into deep water. They ought never to hesitate to cross a stream on a narrow unsafe plank *for fear of a ducking*. They ought never to decline to climb up a tree to pull off fruit, merely because there is a *possibility* of their falling off and breaking their necks. I firmly believe that boys were intended to encounter all kinds of risks in order to prepare them to meet and grapple with the risks and dangers incident to man's career with cool, cautious self-possession, a self-possession founded on experimental knowledge of the character and powers of their own spirits and muscles.
>
> (Ballantyne 1861c: 65)

Thus boyish actions, which are possible (and seemingly necessary) in the unbounded settings of adventure, define and build 'spirits and muscles' – muscular Christians.

The setting of *The Young Fur Traders* is a wild, 'primitive' space, in which the hero's manhood is defined – physically, spiritually and racially. According to his father, Frank Kennedy, Charley would face 'a miserable sort of Robinson Crusoe life, with red Indians and starvation constantly staring [him] in the face' in the HBC Territories. Ballantyne's descriptions of the region are replete with images of the 'primitive' – including 'wild beasts and wild men' in 'grand and savage' landscapes (Ballantyne 1856: 24, 171). 'Wild beasts' include wolves, buffalo and bears, all of which die at the hands of white hunters, but not before they put up a fight, endangering the adventurers. One bear, for example, 'rose, shook himself, gave a yell of anger on beholding his enemies', Harry and Jacques, who respond with a well-aimed, fatal bullet (Ballantyne 1856: 238).[68] 'Wild men' include Indians (aboriginal North Americans), such as one 'savage' who rejects Christianity and retains his traditionally 'cruel' ways, at one point murdering an innocent woman, at another shooting the manager of a fur-trading post. Occasionally, the white boys appear a little savage themselves. Charley, for example, finding his freedom for the first time, 'uttered a shout that threw completely into the shade the loudest war-whoop that was ever uttered by the brazen lungs of the wildest savage between Hudson's Bay and Oregon' (Ballantyne 1856: 43). On horseback, as in snow-shoes, with dog sleds and in canoes, the boy runs wild and free. But he is 'savage' only in play; Ballantyne is very clear about this. Charley is able to play with wildness and savagery, playfully to transgress social conventions and rules, because he is far from his 'civilised' home, in a kind of playground, a space of pleasure but also a space in which identities are dissolved and constructed. *White* masculinity cannot be mistaken for its 'savage' others. The white manliness of Charley, Harry and Hamilton is defined partly in relation to the otherness of *non-white* men; here, masculinity and race are mutually constitutive (see Jackson 1994; Parker 1991; Ware 1992). Hamilton, for example, proves his white manliness by violently punching an Indian who is cruel to the dogs. Hamilton's 'pluck' 'amazes' his companion, Harry, who revises his earlier impression of the former, whom he had once seen as a 'soft' and unmanly fellow (Ballantyne 1856: 219). The Indian attacked by Hamilton, like the bears and wolves killed by Charley and Harry, is an image of 'primitive' otherness, in relation to which white manliness is defined.

The setting of Charley's and Harry's adventures is an almost exclusively male space, and this enables Ballantyne to present their masculinity as if it were independent of femininity and women. The Red River settlement,[69] where *The Young Fur Traders* begins, is a place of girls and women, while the wilderness beyond (including the HBC forts and outposts) is populated almost exclusively by men. While Kate remains at home, 'overjoyed at the thought of being a help and comfort to her old father and mother' (Ballantyne 1856: 15), Charley ventures into the territory beyond. When a *voyageur* (named Louis) considers giving up his adventurous life and returning to the settlement, his 'old woman' calls him 'an old woman', changing his mind (Ballantyne 1856: 61). Women are rarely seen outside the settlement, where most of the action takes place, and when they do appear they are passive and mostly silent. Women – such as Wabisca, who provides a source of tension between Redfeather and Misconna – are variously killed, fought over and saved. They are vehicles through which men define themselves. For the most part, though, women play little or no part in the adventure, which is essentially an all-male affair.

Charley seems to learn everything he needs to know about manhood from men. In the geography of adventure, as in other all-male spaces, a great variety of relationships are channelled between males, and contribute to the construction of masculinities. Relationships of friendship, love and mentorship, between the male characters in Ballantyne's tale, play important roles in the construction of masculinity. Ambiguous relationships of intimate, loving friendship exist between Charley and Harry, as they did between 'real' apprentices such as Ballantyne and his companions. Charley and Harry, for example, are very intimate with each other. When they meet unexpectedly one day, for example,

60

Harry suddenly threw his arms round his friend's neck. As both boys were rather fond of using their muscles violently, the embrace became a wrestle, which caused them to threaten complete destruction to the fire as they staggered in front of it.

(Ballantyne 1856: 80)

Later that evening Charley 'resolved to abandon his own tent and Mr Park's society, and sleep with his friend' (Ballantyne 1856: 81). Relationships such as this, ambiguous as it is, are essential to the boys' rites of passage. They watch and encourage each other's growth, and they also pay close attention to the older *voyageurs*. The boys' 'interest' in the men is equally ambiguous, to the late twentieth-century reader at least, with homoerotic undertones, although the relationship is principally one of mentorship, the boys learning from and emulating the men. The *voyageurs* were, Ballantyne enthuses,

> as fine a set of picturesque manly fellows as one could *desire to see*. Their mode of life rendered them healthy, hardy, and good humoured, with a strong dash of recklessness – perhaps too much of it – in some of the younger men.
>
> (Ballantyne 1856: 55, my emphasis)

While Ballantyne celebrates the muscular physiques and lively spirits of the young *voyageurs*, he reserves his strongest praise for the

> men of middle age – with all the energy, and muscle, and bone of youth, but without its swaggering hilarity – men whose powers and nerves had been tried over and over again amid the stirring scenes of the *voyageur's* life ... they composed a sterling band, of which every man was a hero.
>
> (Ballantyne 1856: 56)

Charley works with, learns from and emulates these men, until he too becomes physically strong and manly in character. His development is so striking that nobody ever seems to recognise him. Harry does not recognise the 17-year-old he meets near Stoney Creek, who is 'tall and stout beyond his years, and deeply sunburnt' (Ballantyne 1856: 117). And Frank Kennedy does not recognise the man who finally returns to Red River, a man who seems 'too big' and even has a 'different nose altogether' (Ballantyne 1856: 268). By then, Charley is a man, physically strong, his rite of passage having taken place in the absence of women.

In *The Young Fur Traders*, the ambiguous, complex relationships between boys and boys, and between boys and men, cannot be sharply categorised, as *either* friendship *or* mentorship *or* love, for example. As Eve Kosofsky Sedgwick argues in *Between Men* (1985), there was no sharp distinction between sexual and non-sexual relationships between men prior to the 'homosocial schism' that took place towards the end of the nineteenth century. There was no need to proscribe intimate, complex relationships between boys and men because these relationships were not necessarily associated with homosexuality (which had yet to be named). The currents of homosocial desire in Ballantyne's tale parallel those identified in classical adventure by Paul Zweig, who observes

> in the literature of adventure: uprooting himself from women, the adventurer forms a masculine friendship so intimate, so passionate, that it reasserts, in male terms, the emotional bond which formerly anchored him within the world of the city. One thinks of Achilles and Patroclus in the Iliad. . . . The adventurer, in his desire to reinvent himself as a man, reinvents his emotions, so that they may be served wholly by male pleasures: the rooted society of women superseded by the mobile society of men.
>
> (Zweig 1974: 75)

Gregory Woods (1995: 136), impatient perhaps with the ambiguity of homosociality, suggests that figures such as Achilles and Patroclus are 'homosexual archetypes', and reads the relationships between modern adventure heroes such as Crusoe and Friday as homosexual. It is tempting to draw the same conclusions of Ballantyne and his characters. Fur traders commonly entered into sexual relationships, not only with native women, but also with each other (Brown 1980; Hyam 1990), and Ballantyne's life-long remorse and penance for the unspecified 'sins' of his youth suggest he was no exception (Nelson 1991). Despite his preoccupation with sex, though, Ballantyne banished explicit sexual acts and sexualities from his adventure fiction. To contemporary readers, the strong currents of homosocial desire flowing between boys and men in *The Young Fur Traders* were not necessarily sexual. In Ballantyne's tale, as in the classical adventures Zweig refers to, masculinity is constructed with reference to homosocial but not homosexual desire, in an apparently self-contained geography of boys and men.

Although the masculine rite of passage seems to take place in a self-contained world of men, it is framed by a series of unacknowledged feminine others. First, the metaphorical femininity of the landscape presents an other against which the hero can define his masculine self. This is a figurative expression of the more general mapping of masculinity (including masculine culture, society) in relation to constructions of femininity (landscape, nature), identified (in American literature) by Annette Kolodny (1975) and, specifically in relation to British colonial geography (in Australian literature), by Kay Schaffer (1988). Kolodny argues that the feminized earth may be constructed as lover, mother or other feminine archetypes. In Canada, Margaret Atwood once generalised, writers often construct nature as 'a woman, but an old, cold, forbidding and possibly vicious one' (Atwood 1972: 200). In *The Young Fur Traders* nature is, indeed, frequently cold, forbidding and vicious. Her 'beauty' and 'mystery' attract men, who 'penetrate' her and find 'perfect paradise' in her 'green bosom' (Ballantyne 1856: 116, 234, 128, 233, 142). But she is fickle, turning cold and treacherous without warning, and claiming men's lives. When adventurers confront this metaphorical femininity, they do so from a safe distance, admiring awesome rather than intimate scenes, and doing so from the detached and commanding security of promontories (see Pratt 1992). They never get very close to nature, even when they have 'penetrated' her. Their relationship to the metaphorically feminine nature, as to the women left behind in the settlement, is one of distance, detachment and difference. They assert their difference to present nature as they do to absent women, the latter more explicitly. One *voyageur*, whom I have mentioned, proves that he is a man by remaining far from his 'old woman'. Another flatters Charley by contrasting him favourably with the less manly fellows who fuss and worry, and are 'as tiresome as settlement girls' (Ballantyne 1856: 131, 151). Masculinity and femininity seem to be poles apart, the latter being banished to literally or metaphorically distant, remote corners. A gulf opens between masculinity and femininity, which become caricatures, dialectically opposed to each other (see Kroetsch 1989).

The construction of masculinity in Ballantyne's story, in relation to absent and obscured femininity, was paralleled in the context in which the book was produced and consumed. Ballantyne's boys' stories, in general, were framed by unacknowledged girls and women. Ballantyne told stories to absent women, defining himself, or other men, to an unacknowledged female muse. His yarns, despite their masculine style, reminiscent of the camp fire and the 'Bachelors' hall' (in HBC forts), were actually recounted in letters to his mother in Scotland. While his mother saved the letters in an embroidered silk wallet, it was a kindly aunt who encouraged Robert to write his first book (a Canadian memoir), and paid for it to be privately published (in 1848), and it was his sisters who marketed the book, ensuring its sale by subscription. Generally speaking, Ballantyne's mother and his sister-in-law were the prime influences on his life as a young writer, while male relatives including his father were marginal and distant figures in Robert's biography (he hardly noticed when his father died, but was devastated by the death of his mother). And while his stories were addressed and marketed to boys, many of Ballantyne's stories and characters, including Ralph Rover (hero of *Coral Island* and *The Gorilla Hunters*), were invented to entertain the writer's two nieces, Dot and Edith, and were very loosely based on the fictional adventures of their brother, Randal (Quayle 1967). And, of course, girls and women read Ballantyne stories, although they were not acknowledged as readers. Unacknowledged female readers of 'boys' stories' were not incidental to the telling of the masculinist adventure; boys and men knew that 'their' stories were being read by women, that 'their' masculinity was recognised by girls and women.

The setting of *The Young Fur Traders* is a simplified, seemingly uncomplicated space, in which a simplified and homogeneous masculinity is made to seem plausible. Whereas the masculine imagery in certain school stories is heterogeneous and relatively subtle, its counterpart in adventure stories is homogeneous and simple. Ballantyne was openly intolerant of 'muffs' – boys who are 'mild, diffident and gentle', and who are 'disinclined to try bold things' (Ballantyne 1861c: 70). He argued that for a boy to become an 'effective' man, able to 'protect a lady from insolence; to guard his house from robbery; or to save his own child should it chance to fall into the water', he must start by learning to swim and run, and must boldly do things like wrestling, '"jinking" with [his] companions' and leaping 'off considerable heights into deep water' (Ballantyne 1861c: 71). In *The Young Fur Traders* the 'muff' is Hamilton (also known as Hammy), a 'soft', 'slender' 19-year-old. Hamilton is not particularly keen to rough it, nor is he particularly keen to become manly. But he is given no choice. When he hesitates to accept Harry's 'invitation' to join a 10-mile snow-shoe tramp in sub-zero temperatures at night, Harry loses patience with him and snaps, 'Come, man, don't be soft; get ready, and go along with us' (Ballantyne 1856: 167). Harry is not satisfied until Hamilton gets frost-bitten in both feet and hobbles bravely on (Plate 3.4). Hamilton's development during his stay at York Factory shows that any male, adventurous or not, can become manly. His most dramatic transformation occurs during a two-week journey with Harry, *en route* to Norway House. During the arduous journey, Harry and Hammy soon become tired and blistered, but they find that after much painful exertion 'their muscles harden, and their sinews grow tough', and they even begin to enjoy the ordeal (Ballantyne 1856: 221). Hammy may always be a sensitive, gentle sort of man, but life in the northwest has brought out his manly qualities. While Ballantyne could argue that there was no place for 'muffs' in Britain, he could quite plausibly show that there was nowhere for a 'muff' to hide in the rugged, frozen wastes of British North America. Ballantyne used the simplified, carica-tured ruggedness of his setting to make a simplified, universal masculinity seem plausible. Simplification like this would surely have seemed excessive, crudely didactic, if it had appeared in any medium other than popular, juvenile literature. One boys' story writer, conscious of the extent to which he simplified, made the excuse that he was writing for boys, claiming that, 'In the endeavour to interest the juvenile intellect, it is necessary to deal with physical rather than moral facts. . . . Show and style have been sacrificed upon the altar of simplicity' (Reid 1853: vi). But juvenile audiences did not force so much as excuse such 'simplicity'. Boys' adventure stories were, after all, consumed by adult as well as juvenile readers, and by youths who also liked Dickens, Scott and other relatively complex authors (Salmon 1888). The juvenile medium did not force Ballantyne to assert and naturalise an extreme ideology of masculinity; it enabled him to do so.

Finally, the setting of *The Young Fur Traders* is a realistic space, in which Ballantyne's ideology of masculinity is naturalised. Although Ballantyne did explicitly assert his right 'to transpose time, place, and circumstance at pleasure' (Ballantyne 1856: 3), he (unlike some of his contemporaries[70]) always attempted to get the basic geographical 'facts' right, and was applauded by critics for his geographically informative writing (Salmon 1888; Yonge 1886). For example, *The Athenaeum* commented of *The Young Fur Traders* that 'the descriptions of hunter-life in the backwoods, and the society and manners at the trading stations of the Hudson's Bay Company, are excellent, and have unmistakeable signs of having been drawn from life' (Quayle 1967: 106). No knowledgeable

Plate 3.4 'One of the evils of snow-shoe walking', an illustration in Ballantyne's *The Young Fur Traders* (1856), based on a drawing by the author. Mild-mannered Hammy suffers frost-bite and exhaustion, much to the amusement of his companions, who are concerned mainly that he behaves 'like a man'. The realistic, rather than freely artistic, portrayal of fur-traders' costumes and weapons serves to naturalise the story and ground its moral lessons in believable space.

fur trader would have contradicted Ballantyne's claim that *portages* varied in length from around 12 yards to around 12 miles, and no British geographer familiar with Arrowsmith maps of British North America would have disputed Ballantyne's general configuration of lakes, rivers and forts. For realistic details, Ballantyne drew partly upon his experiences with the HBC (at York Factory, Fort Garry and other places described in the book), when he observed the ways and the settings of the fur trade, and listened to the yarns told in the fur trading posts and around camp fires (Peel 1968; Quayle 1967, 1968; Selby 1963). His observations also informed the realistic, 'illustrative wood-cuts' he 'made on the spot' (Ballantyne 1848: xi), which appeared in his memoirs and in early editions of his novels. Ballantyne's images emphasise detail and portray life in the Canadian North West faithfully, in a manner that is informative and educational. For a man described as an 'accomplished artist' (Whalley and Chester 1988: 134), they are decidedly artless. Ballantyne's illustration of 'one of the evils of snow-shoe walking' (Plate 3.4), for example, depicts in faithful, educationally realistic detail, the clothes and footwear of the male figures, the equipment they carry and their snow-shoe technique – outstretched arms preserving the balance of the two who remain standing. Despite this attention to detail, in descriptions and illustrations, Ballantyne did not 'teach' geography or natural history as an end in itself. Images of geography and natural history served to code the real – moral – lessons of *The Young Fur Traders*. The realistic imagery presented a believable world, in which an ideology of masculinity was grounded.

CONCLUSION

The geography of adventure, cultural space charted by adventure stories such as *The Young Fur Traders*, accommodates and conditions constructions of identity. The contours of this cultural space do not determine, but nevertheless set parameters around, the identities they accommodate. Men are made, albeit loosely, in the image of their settings. While they appear to be set free in the geography of adventure, they are also confined to the limited range of masculine identities that are possible there – broadly speaking, they are confined to hegemonic masculinity. As images of hegemonic masculinity (Donaldson 1993), Ballantyne's boys and men are heroic but 'ordinary'; they are stock heroes, familiar and even 'normal' to readers of popular literature; and they are literally and metaphorically distant from and in fearful awe of women and femininity.

The boys and men mapped in adventure stories are *dramatis personae*, literary heroes rather than real people, so they and their masculine identities should not be taken too literally. Although they can sometimes be mistaken for ordinary, real boys and men, and can therefore naturalise the masculinity they embody, adventure heroes like Charley are simple figures with none of the complexity and multi-dimensionality of real people. Somehow they do map onto real gendered subjects. As one boys' magazine reader recently reflected, 'boys' comics, in particular, played a small but significant part in the ideological construction of my masculinity' (Jackson 1990: 223). But, as Graham Dawson illustrates in *Soldier Heroes* (1994), his partly autobiographical study of adventure and masculinity, relationships between readers and heroes are complex. To assume that, as one contemporary critic put it, 'the *dramatis personae* of a story become living entities', would be to oversimplify the relation between boy readers and boys' stories, and to ignore all other readers (Salmon 1888: 209). The only known cases of *dramatis personae* becoming real entities were the authors themselves: Bracebridge Hemyng assumed the identity of Jack Harkaway (Turner 1957), while Ballantyne was widely known as 'the brave Mr Ballantyne' (Quayle 1967: 103–104), and Henty presented himself as a role model for boy–men (Reynolds 1990).

The power of a geographical narrative rests partly on its ability to block other narratives, other maps of geography and identity (Said 1993). Adventure stories such as *The Young Fur Traders* blocked alternative constructions of masculinity, for example. One observer wrote, as late as 1961, that Ballantyne's stories provided him with the 'only map' of his 'particular reality' (Niemeyer 1961: 243). Not only was Ballantyne's ideology of masculinity accepted (read) by boys who could directly emulate the heroes, it was also tacitly accepted by those who could not – boys who were not white, heterosexual, middle-class, British, English-speaking, Christian – and by girls and women. These readers were marginalised through their acts of reading stories that excluded them and legitimated the 'natural' authority of others. If they were willing and enthusiastic readers of such stories, as Turner (1957) suggests they were, that was partly because they were denied stories of their own, in which they would be acknowledged as readers, and in which they could identify directly with the heroes.[71] This marginalisation and exclusion helped to underpin the hegemony of the male élite.

In this chapter I have emphasised ways in which masculinist geographical narratives map geographies and identities, although I do not wish to understate the agency of readers, nor their abilities to negotiate critically and creatively the geography of adventure and its maps of identity. While hegemonic masculinity becomes taken for granted, naturalised in everyday practices such as the consumption (including reading) of masculinist geographical narratives such as adventure stories, the corollary of this is that it can be denaturalised in the same sorts of practices, the same cultural spaces. Ballantyne's geography, replete with inversions and disorder, has the potential to be a site of resistance, in which writers and readers may subvert powerful constructions of geography and identity; this story I pick up below. First, I turn to the second part of my initial claim, that adventures map masculinities in relation to geography, and geography in relation to masculinities, by refocusing attention from metaphorical to material spaces.

4

MAPPING EMPIRE
Space for boyish men and manly boys in
the Australian interior

Territory and possessions are at stake, geography and power.

(Said 1993: 5)

Edward Said reminds us that the geography of adventure is not all metaphorical. Adventurous imaginations not only appropriated metaphorical spaces, mapping European identities in non-European settings, they also appropriated material spaces, mapping and making European colonies in the non-European world. As they engaged popular geographical imaginations, adventure stories promoted popular support for, and involvement in, imperialism. Anthropologist Andrew Lang observed, in 1891, that 'men of imagination and literary skill have been the new conquerors, the Corteses and Balboas of India, Africa, Australia, Japan and the isles of the southern seas' (Lang 1891: 198). But novels and other forms of colonial culture did not 'cause' colonial acts; the writing and reading of adventures and other novels, both in Europe and in the colonies themselves, were part of the process of (ideological and material) colonialism. Adventures constructed imaginative space in which colonisation could take place, and sometimes mapped the course of that colonisation.

Adventure has been labelled an imperialist literature, often justly, although relationships between adventure stories and geographies of empire are many and varied. Adventures map colonial spaces at a variety of geographic scales, ranging from the generic space of *Robinson Crusoe's* island, as it appeared in nineteenth-century abridgements, to the more specific – South and Central American – colonial geography of Defoe's original novel, to the localised settings of adventure stories such as Traill's *Canadian Crusoes* – on the Rice Lake Plains of Ontario. Adventures vary, too, in the forms of imperialism they represent. Early modern adventure literature was associated with mercantile capitalism, while *Robinson Crusoe* marked a shift towards *petit bourgeois* colonialism, founded more on emigration, settlement and practical work (Nerlich 1987). In the Victorian period alone, some adventurers went in search of gold, while others went in search of land to till, others still in pursuit of entirely different imperial dreams (Moyles and Owram 1988). Some adventure writers and stories were directly and explicitly imperial, others indirectly and implicitly. While some merely failed to

68

stand in the way of empire (Said 1993), others were active empire builders, demonstrating commitment to the imperial cause in all areas of their lives. Adventure writers from Defoe in the early eighteenth century to Haggard, Ballantyne, Kingston and others in the late nineteenth, promoted British imperialism not only in their writing, but also in their campaigns for overseas investment and emigration, especially to Canada, Australia and other white settlement colonies. While they regarded themselves as patriotic, writers of adventure fiction did not simply promote and protect a pre-defined empire. Through their literary work, they sought to play a part in building the empire, in sculpting and shaping it, according to their own ideas, some of which were 'progressive'. Thus the imperial geographies of adventure literature were not only as diverse as those 'on the ground', they were as diverse as opinions and visions of imperialisms of the future. There are therefore no easy generalisations to be made on the imperialism of adventure. It is possible to say, however, that particular historical forms of adventure chart, and are mirrored in, particular historical geographies and forms of imperialism.

Colonial geographies reflect the characteristics – including the masculinity – of the geographical imaginations in which they were conceived, the geographical narratives such as adventure in which they were mapped. The 'new conquerors' who wrote imperial adventures were, in the words of Lang, specifically '*men* of imagination and literary skill' (my emphasis). The masculinism of adventure stories – almost all of which are by men, for boys and men, and about boys and men – is imprinted on the geographies they map. The identity of a 'young man' – a white, unmarried colonist such as Robinson Crusoe – is stamped on empire, frontier and exploration (Green 1979). But there is no universal young (and/or single) man, even though Crusoe has sometimes been interpreted as such. Geographies of adventure bear the imprint not of archetypal masculinity, not of generic 'young men', but of historical masculinities. The geography of Ballantyne's mid-century Canadian adventure, for example, bears the imprint of British Victorian hegemonic masculinity, specifically Christian manliness. In this chapter I demonstrate, specifically, that the colonial geographies of late nineteenth-century adventure stories, set in actual or possible colonies, bear the imprint of late nineteenth-century masculinities.

Between mid-century, when Ballantyne's first books were published, and the final decades of the century, when a particularly active phase of British imperial expansion (in the form of territorial acquisition) was in full swing, adventure literature generally changed. In particular, adventure became more closely and directly imperial in the late-Victorian and Edwardian periods. The British Empire expanded rapidly in the period between 1870 and 1914, not only through the colonial acts of Britain's élite, such as explorers, sportsmen, investors and military leaders, but also through the colonial acts of British common people. More than ever, popular literature was directly connected to colonisation, not just representing or legitimating colonisation. And adventure stories were often explicit and specific in their promotion of colonialism.

In the context of market growth in literature and consolidation of popular literacy, which approached universal levels around the turn of the century, adventure stories became more secular and more violent, and in the context of evolutionary theory they became more racist. Lively, Christian stories were generally supplanted by the theologically vague works of writers such as Robert Louis Stevenson and Rider Haggard, in which fun triumphed over faith. Violence reached new levels in stories such as *Treasure Island* (1883) and *King Solomon's Mines* (1885). Never before, in respectable Victorian literature, was violence so graphic, gratuitous and lighthearted, so calculated to entertain. Violent stories such as these, while frequently condemned by many critics and educators, were commercially successful (A. White 1993). Violence was commonly inter-racial, its victims non-whites. Victorian ideas about race, guided by Darwinian evolutionary theory, filtered through to adventure stories in settings and stories in which whites encounter stock 'savages' and members of 'degenerate' races. Clashes between white heroes and non-white beliefs, customs and bodies generally left whites in better shape and confirmed white superiority.

The new brand of racist, imperialist, masculinist and violent adventure literature mapped British colonies, including settlement colonies. Haggard's fiction promoted British colonisation in central Africa without offering popular readerships much opportunity for involvement as independent colonists (although whites did emigrate to southern Africa). But adventure stories of the same brand, often modelled directly on Haggard, also presented readers with more concrete colonial opportunities. One such story, completed by British-Australian colonist Ernest Favenc in Sydney in 1894, was titled *The Secret of the Australian Desert* (1896a).

The Secret follows *King Solomon's Mines* closely, substituting Australian settings and characters. Favenc's heroes, despite their many ancestors, are most closely related to Sir Henry Curtis, Captain John Good and Allan Quartermain, in *King Solomon's Mines*. Haggard's three adventurers travel into unmapped African desert in search of a lost Englishman, Sir Henry's brother George. They find a lost kingdom, and the legendary source of King Solomon's riches. They endure the heat of the desert, survive encounters with fierce warriors, and tangle with seemingly evil practices and evil spirits in the course of their exciting adventures. Their mission accomplished, they leave the interior, laden with diamonds. Similarly, in *The Secret*, three white British men travel through the Australian desert in search of a legendary burning mountain and a lost explorer. They endure the heat of the desert and triumph over cannibals, before leaving the interior, laden with gold. Favenc's tale, like Haggard's, imaginatively revitalises the British race, and specifically British men, in the space of adventure. Favenc departs from Haggard by connecting this process of revitalisation to colonisation through white settlement.

The biography of Favenc – colonist, adventurer and writer – illustrates some of the connections between adventure literature and imperialism. Favenc emigrated from Britain to Australia in his early twenties. Once there he worked on a remote Queensland cattle station, participated in brutal aboriginal 'dispersals', was involved in mineral prospecting, and tried his luck as an explorer. Favenc's biography reads like an imperial adventure story, in which the hero repeatedly strikes out into the unknown, where he fights off hostile natives and endures terrible drought, before returning to civilised society and telling his story (Frost 1983). All 'great adventurers', it is said, 'have not only been great doers, they have been great talkers' (Zweig 1974: 81). Favenc was both a doer and a talker, although he was more prolific and successful as a talker, a teller of tales. He loved exploration and he worshipped the 'great' Australian explorers, but he was born too late to stand much chance of being a 'great' explorer himself. He arrived in Australia in 1863, just after Burke and Wills had perished, and fifteen years after Leichhardt had gone missing. European Australians seemed to be exploring the continent's few remaining *terrae incognitae*, and the age of heroic inland exploration was drawing to a close. Favenc's career as an explorer, first as an amateur, later as a professional (he was commissioned to explore the route for a proposed railway in Queensland), was short and not very successful; as an explorer, he was never to win much fame or recognition. For the most part he had to content himself with retelling and reworking the stories of other explorers, generally the explorers of a bygone age. So he settled down in Sydney to write, mostly about Australian exploration.

Favenc's writings, which range from relatively formal histories and geographies to works of poetry and adventure/mystery stories, exhibit a general preoccupation with geographies of Australian exploration. His most famous work remains *The History of Australian Exploration from 1788 to 1888* (1888), which was published in British Australia's centennial year, and was to be regarded as the definitive history of Australian exploration for at least half a century (Frost 1983). Explicitly colonialist, if not nationalist, *The History* was dedicated to the premier of New South Wales, 'the mother colony, from whence first started those explorations, by land and sea, which have resulted in throwing open to the nations of the world a new continent, now rapidly developing' (Favenc 1888: iii). *The History* chronicles the first phase of British-Australian inland exploration and colonisation, when colonists explored regions in which to extend their colonies. Many of the sentiments and themes of *The History* – its British-Australian patriotism, its emphasis on adventurous but sober exploration, its preoccupation with the interior – are seen in Favenc's other works. These include the non-fiction histories *The Explorers of Australia, and their Life-work* (1908) and *The Great Austral Plain, its Past, Present and Future* (1881), and the non-fiction geographies *The New Standard Geography of Australasia* (1898) and *The Geographical Development of Australia* (1902), a school textbook. Favenc also wrote juvenile adventure stories set in and around Australia, including many short stories, and three full-length novels, *The Secret of the Australian Desert* (1896a), *The Moccasins of Silence* (1896c) and *Marooned on Australia; the Narrative by Diedrich Buys of his Discoveries and Exploits in Terra Australis Incognita about the Year 1630* (1896b). The diversity of Favenc's exploration narratives mirrors and illustrates that of

Australian exploration literature as a whole. His exploration adventures, of which *The Secret* is broadly representative, mirror and illustrate contemporary Australian exploration adventure literature as a whole.[72]

The Secret of the Australian Desert is typical of the many individually obscure retellings of a British-Australian imperial narrative – the inland exploration adventure. Like the many adventure yarns retold and invented by colonists in trading posts, around camp fires and in make-shift beer parlours around the British Empire, Favenc's fictional stories have faded into general obscurity, and have been outlived by their canonised counterparts. But, like many other retellings that have since become obscure (including some, very similar, by writers such as John Mackie and Alexander Macdonald), Favenc's story helped map a British colony and chart its future. The colonial map Favenc constructed reflected the characteristics – including the particular, historical masculinism – of contemporary adventure fiction.

READING AND RETELLING ADVENTURE STORIES IN COLONIAL AUSTRALIA

Reading adventure stories in British colonies was part of the process of mapping and making those colonies. While Britons read stories set around the British Empire, participating in the imaginative mapping of that empire, British imperial culture was not confined to the shores of Britain. British books, newspapers and magazines were distributed (by publishers and private correspondents), purchased and read around the world, particularly around the empire. In settler societies, such as Canada and Australia, literacy rates among British settlers were generally at least as high as in Britain. At the turn of the century, 95 per cent of adult Australians could read and write, and Australia could be described as 'a reading-oriented society' (Lyons and Taksa 1992). Colonists were commonly avid readers, although some, particularly those in rural areas, found it difficult to obtain reading material. Many emigrants included books in their luggage (Tulloch 1959), and once settled they received books by post, mainly from friends and relatives, and from organisations such as the Lady Aberdeen Society, which provided some isolated colonists with reading material. They also borrowed books, mostly from each other, and eventually from libraries (Mein 1985).

Most of what colonial Australians and Canadians read was published in Britain. As Australian juvenile literature specialist Brenda Niall has observed, 'Until the 1950s, Australian children's reading came from the same shelf', and often the same publishers, 'as that of their English contemporaries' (Niall 1988: 547). Prominent among British-produced books and magazines were adventure stories, which were as popular in British colonies and dominions as they were in Britain itself. A survey of reading among Canadian pioneers identified the popularity of British adventure story writers such as Stevenson and Henty (Tulloch 1959). And a recent oral history confirms the popularity of British adventure books and magazines in colonial Australia (Lyons and Taksa 1992).

Among the most popular books and magazines were *Robinson Crusoe* and *Treasure Island*, and *The Boy's Own Paper*, respectively. A catalogue of juvenile literature (Angus and Robertson junior book club catalogue) published in Sydney in 1907 listed many of the same titles that were popular in contemporary Britain, including sixty Ballantynes, fifty-five Hentys and forty-one Kingstons (Lyons and Taksa 1992: 8, 94). Some of these British-made books were set in the colonies. W.H.G. Kingston, for example, added *Australian Adventures* (1884a) to a series which included such titles as *Adventures in Africa* (1883), *Arctic Adventures* (1882), *Adventures in India* (1884b) and *Adventures in the Far West* (1881).

But colonials did not just read purely British literary products. In some cases there was colonial input in the 'British' stories. Some British-born writers travelled, worked or lived for a time in the places where they set their stories, which were generally published in Britain. For example, Ballantyne worked for five years in Canada, where he gathered material for the stories (including thirteen books) he wrote later in life, set in Canada but written and published in Edinburgh. John Mackie, whose stories include *The Heart of the Prairie* (1901) (set in western Canada) and *The Lost Explorer* (1912) (set in Australia), had been a mounted police officer in Canada and a mineral prospector in Australia. Ernest Favenc, as I have mentioned, was both colonist and writer, and he wrote for publishers and readers in both Australia and Britain. His career illustrates how the boundaries between British and Australian colonial literature are straddled by 'Australiana', cultural products with Australian content that are not always recognised or claimed in Australian literary histories as Australian literature. But, whether they are labelled British, Australiana or Australian, colonial cultural products such as adventure stories that represented Australia generally incorporated a mixture of British and Australian content.[73]

In many cases, colonial cultural producers followed established British traditions and styles, sometimes adding local content. 'High' colonial arts, including painting and poetry, were often reminiscent of their British counterparts (although, by contemporary critical standards, they were generally less accomplished). And adventure stories set or written in British colonies were often retellings of earlier British narratives, with local settings and some local characters (such as explorers and bush rangers). For example, Traill's *Canadian Crusoes* (1852) was a Canadian colonist's version of a generic island story (*Robinson Crusoe*), while Favenc's *The Secret* was an Australian colonist's version of an African story. Like many other Australian writers, Favenc retold and reworked stories about journeys 'into the unknown': stories of adventure and exploration. Explorers were a favourite topic of many British adventure story writers and readers. Many of the most popular Victorian adventure storytellers turned to explorer themes. Writers such as Kingston and Ballantyne, who retold contemporary exploration stories or used exploration stories as points of departure for fiction, provided the basic formulae and cast of characters for Australian exploration adventure stories.[74] Australian exploration adventure stories reflect a general interest in explorers, shared by readers around the British Empire, and also a specifically Australian fascination with exploration and with the Australian continental interior.

In his adventure and exploration stories, Favenc picked up on a fascination with the 'unknown', which he felt himself and sensed in his fellow colonists – Europeans attempting to possess and inhabit vast areas of what, to them, was *terra incognita*. Through the Victorian period, many Australian colonists shared the explorers' interest, preoccupation even, in the mystery of what lay between their colonies, in the continental interior. Until the middle of the nineteenth century, most Australian colonists had lived in pockets of concentrated and, by contemporary standards, highly urbanised settlement along the coast, metaphorical islands separated by oceans of unknown space in a metaphorical Australian archipelago.[75] 'If they stepped outside' their little colonies, 'they were lost' (Moorehead 1963: 10). As one popular historian has put it, 'the excitement of venturing into the unknown . . . was never absent from the public imagination in Australia's "Furious Fifties"' (Clune 1937: 4). Australian colonists were preoccupied with the 'perennial' problem of 'what was *there*, at the heart of the continent, in the "ghastly blank" in the centre?' (Serle 1973: 7). This preoccupation was manifest in fascination with explorers, notably Ludwig Leichhardt, who disappeared without a trace in 1848.[76]

Between the time of Leichhardt's disappearance and the end of the nineteenth century, vast regions of the Australian continent were colonised. Explorers, overlanders and other travellers forged land connections between the colonies, and by the early 1890s most of the potentially useful[77] land in Victoria, New South Wales and Queensland was explored and taken up, usually by cattle ranchers (such as Favenc) then sheep ranchers.[78] But, as late as the 1880s and 1890s the Australian interior remained, both physically and imaginatively, a largely 'unknown' and particularly 'debatable' landscape (in British-Australian geographical imaginations). Inland colonisation was rapid, but it was also hit-or-miss. South Australians colonised large areas to the north of their existing settlement, only to be pushed back by the drought of 1880–1884 (Meinig 1962). Squatters in Queensland and New South Wales helped increase Australia's sheep and cattle populations to 106 million and 12 million respectively in 1891, but they too encountered drought, which reduced their stock levels by a half between 1895 and 1902. Setbacks in some areas were matched by advances in others. For example, dramatic progress was made in wheat production, which increased tenfold in the last four decades of the century, and in artesian water production, which peaked in 1900. As colonists rushed into Australia's continental interior, there was a great deal of uncertainty about what that land was, and what it could be. Some people believed the continent could support from 1 to 500 million people in the future, while others thought it was a desert, and would always be a desert (Powell 1988). Some (such as Henry Lawson) found the interior monotonous and ugly, while others (notably A.G. Stephens) claimed to see beauty in it (Barnes 1986). The Australian interior, a complete mystery to most white

Australians in the middle of the century, was still something of a mystery as the end of the century approached. Perhaps that was why stories of explorers who entered the mysterious interior continued to fascinate Australians. No explorers fascinated them more than Leichhardt and Burke, who died *en route*. Also celebrated were Sturt and Eyre, who returned with no useful discoveries to report, only stories of daring and suffering.[79]

Ironically perhaps, the tragic and heroic explorers, who most fascinated Australians, were generally excluded from or marginalised in formal histories and geographies of Australian exploration. Formal narratives of Australian exploration, like those of inland exploration in other continents, particularly North America, emphasise epic quests. Typically, in *The History of Australian Exploration from 1788 to 1888*, Favenc tells the stories of explorers who go out into 'unknown lands' where they face 'hardship and danger' and display 'courage and fortitude', before successfully attaining their goal(s), whereupon they 'return to civilized communities with the tidings' (Favenc 1888: vi, 17).[80] The explorers who best fit this pattern were early nineteenth-century figures including George William Evans, the first European to cross the Blue Mountains (1813), and John Oxley, Charles Sturt and Thomas Mitchell, who explored further inland. Some other explorers who did not fit Favenc's narrative very naturally were forced into it, made to seem more heroic or successful than they really were. For example, some explorers who thought they saw deserts were suffering tiredness and/or illness, or were travelling in a time of drought.[81] Others whose stories could not be told as heroic quests were left out of *The History*. These include private travellers, such as squatters and overlanders, who did 'the bulk of the detail work' in exploration, but whose stories would 'prove most monotonous reading, and fill, I am afraid to think, how many volumes' (Favenc 1888: v). Also excluded were Aborigines, women travellers and a number of male explorers, notably Eyre, Burke and Leichhardt, who did not fit the mould of the epic explorer, either because they were insufficiently heroic, or because they were too heroic – insufficiently practical or egotistic. In Burke, for example, Favenc could not 'help being struck by the exaggerated and misplaced stress laid upon the reputation Burke possessed for personal bravery' (Favenc 1888: 210). Similarly, Eyre was courageous, to the point of self-indulgence. His obsession with 'being the first white man to cross the desert' was, in Favenc's judgement, inexcusable (Favenc 1888: 132). 'As for any knowledge of the interior that was gained, of course there was none, even the conjectures of a worn-out, starving man, picking his way painfully around the sea shore, would have scarcely been of much value' (Favenc 1888: 136).

Favenc recognised that the interior desert, while unsuitable as the setting for conventionally epic exploration history, had some potential as a site, whether material or metaphorical, for Australian colonialism. So, having bypassed most of the interior in *The History*, he returned there in other works, and rehabilitated explorers whom he had recently dismissed. Favenc acknowledged that Burke and Leichhardt had caught Australian imaginations, and had some part in Australian history. Burke, he observed, was 'elevated into a hero' (Favenc 1888: 208), while Leichhardt's disappearance was 'one of the strangest mysteries of our mysterious interior' (Favenc 1888: 166). Burke and Leichhardt were the raw material for a different kind of imperial history. Writing this history, Favenc turned to freer language, to write freer histories and geographies. He wrote romances in which he took some 'liberty with history' and admitted to enjoying 'some freedom with chronology', trusting 'that in a romance the inaccuracies will be pardoned' (Favenc 1896b: v–vi). Favenc used adventure stories as a way of exploring the ambiguities and mysteries of lost explorers and unknown landscapes. In *The Secret*, he took up where he left off in *The History*, reworking the story of Leichhardt's final expedition and his mysterious disappearance, and charting the mysterious geography of central Australia.[82]

MASCULINIST NARRATIVE AND COLONIAL GEOGRAPHY

In *The Secret*, the Australian interior is the colonial setting of a masculinist adventure. Although representations of masculinity are prominent, the story is less about men than it is about empire. Unlike many British-based writers, such as Haggard and Ballantyne, Favenc was a colonist, less concerned with (British) manhood than with (Australian) colonisation. Above all, the colonial space charted in *The Secret* accommodates the colonial dreams of white men, conventionally masculine figures. These men embody a late nineteenth-century variation on the theme of Christian manliness. Boys and men, in the exclusive company of boys and men, they are muscular, fun-loving, and motivated at least partly by the love of adventure and excitement. They are anti-intellectual, while knowledgeable about bushcraft, survival skills and subjects such as geography and botany. They are sometimes ambivalent about their Christian faith, but never about their racial superiority. Neither are they ambivalent about their colonial ambitions, even though these are largely selfish and frequently violent. The geography of *The Secret* is a projection of these masculinist colonial desires.

A dualistic geography of adventure, in which home and away are kept separate, the setting of *The Secret* is off the British-Australian map and away from home. The adventure is conceived at the home of a colonist, Morton, who owns a cattle station in South Australia, where he sits smoking with his friend, Brown, and his young nephew, Charlie, on the back porch. The three gaze away into the distance, just as other prospective heroes gaze at maps, before setting out on their adventures. Lured by the promise and the possibility of 'unknown' space, they resolve to leave the cattle station in search of a legendary 'burning mountain' somewhere in the Australian interior. Next morning their quest begins. 'With full water bags, and a determination not to be beaten back' (Favenc 1896a: 15), they head for the interior, the mountain their vague destination. Their passage, off the edge of the map and into the geography of their adventure, takes the form of a journey through forbidding, stony desert, a 'bare expanse of rock' (Favenc 1896a: 48) where all tracks seem to tend towards some unknown centre – the mythical centre of the Australian continent. As the heroes pass from known to unknown space, the landscape fades from detailed, textbook geography to clichéd Australian literary landscape – weird, melancholic, silent, strange and sullen – as the three get further from home. Favenc's sublime interior is reminiscent of the 'Weird Melancholy' observed by Marcus Clarke in 1876, in which 'All is fear inspiring and gloomy'. Clarke wrote how 'Hopeless explorers have named [mountains] out of their sufferings – Mount Misery, Mount Dreadful, Mount Despair. . . . In Australia alone is to be found the Grotesque, the Weird, the strange scribblings of Nature learning how to write' (Clarke 1876: v–vi). In *The Secret*, similarly,

> It was a weird and weary tramp across this rock by the light of the stars, with vague darkness all around them. None of them felt inclined to speak, and an intense silence reigned everywhere. A sickly moon rose just before daylight, and its faint beams cast the long shadows of the travellers across the gleaming surface of the limestone.
>
> (Favenc 1896a: 47)

Beyond the limestone plain is a volcanic landscape, the least familiar, least humanised space of all. Together, these landscapes are clearly alien to the heroes, who now find themselves off the map, in territory they neither know nor understand. The trio pass into the geography of their adventure, the liminal terrain in which masculine rites of passage take place, and the unknown terrain in which colonial desires are accommodated. Vague and generic landscapes, not necessarily betraying geographical ignorance, or what art historian Bernard Smith once called 'European vision', create cultural space in which to invent colonial Australian geography.

As a plausible space in which constructions of masculinity and imperialism are naturalised, the setting of Favenc's adventure story is realistic. Favenc's story is grounded in believable geography, less detailed than the geography he described in more formal history and geography books and textbooks, but otherwise not too dissimilar, particularly around the relatively well-mapped edges of the interior. Favenc explicitly attempted to get basic geographical 'facts' right. He assured readers that 'descriptions of the physical features of the country are faithful records from personal experience' (Favenc 1896a: v). *The Secret* begins in plausible space, with details of bush flora and fauna, of bushcraft, of practical exploration, all drawn (in part) from Favenc's personal experience. He describes local vegetation such as bloodwood trees and mulga scrub, and uses specifically Australian terms such as 'gins' and 'pickaninnies' to refer to Aborigines (women and children). He describes bushcraft techniques, such as locating water holes by observing the 'flights of white corellas' (Favenc 1896a: 19). Realistic geographical detail thins out as the heroes enter the interior, although the realistic appearance of *The Secret* is not compromised, since the interior is a hazy region on the contemporary British-Australian mental map. Realistic, if not well known, the geography of the interior is solid ground upon which to rehearse an ideology of masculinity, and also upon which to build dreams of a colonial future.

The setting of *The Secret* is a space of mystery and legend. The realistic geography that frames the 'unknown heart of the continent' only serves to emphasise the evasiveness and mystery of the interior. In *The Secret*, Favenc emphasises and perhaps exaggerates the mystery of central Australia, as comparison between the maps in his formal exploration history and his boys' exploration adventure story make clear (see Plates 4.1, 4.2). The fold-out map included in *The History* is crammed with detail, presenting the interior as a land much-traversed and relatively well known, whereas that in *The Secret* is sketchy and virtually empty, its white space constructing – rather than representing – *terra incognita* (see Shohat 1991). The *terra incognita* of *The Secret* is marked with the route that may have been followed by Leichhardt, a tragic figure who passed through the region without conquering it. Leichhardt's disappearance defines the desert as the setting of a great mystery. So does the legend of the 'burning mountain', which Morton, Charlie and Brown initially set out to find. The desert is replete with secrets, mysteries and riddles, which are manifest in cave

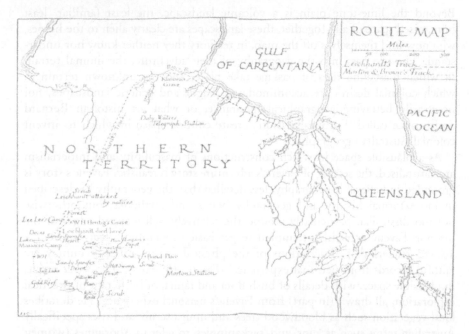

Plate 4.1 The frontispiece to *The Secret of the Australian Desert* (1896). This route map, which shows 'Leichhardt's Track' and 'Morton and Brown's Track', grounds Favenc's fiction in realistic space, bringing the fictional characters Morton and Brown together with historical figures such as Leichhardt in 'real' but largely unmapped space. The land is coloured sandy yellow, the ocean light blue.

Plate 4.2 The fold-out map included in *The History of Australian Exploration from 1788 to 1888* (1888), crammed with detail, contrasts sharply with the literary and cartographic maps in Favenc's adventure stories.

paintings that seem to suggest the presence of an 'ancient and partly civilized race' (Favenc 1896a: v), and in number of other unexplained artefacts and strange happenings.[83] The adventurers find mysterious artifacts, some of which seem to be clues: a dead native with a red smear on his forehead and a white triangle painted on his chest; a six-toed footprint ('the footprint of the devil') carved on a rock surface; an anchor carved on a tree; and, living with cannibals, an 'old, old man, with snow-white hair and beard' (Favenc 1896a: 62). All these clues present new mysteries, some of which the heroes unravel. For example, it turns out that the carved anchor was left by an illiterate white man, a member of Leichhardt's party. The old white man, living with the cannibals, proves to be a survivor of Leichhardt's party. The rather mutilated copy of a journal, written by another of Leichhardt's party, finally resolves the mystery of Leichhardt's disappearance (Plate 4.3). But while the adventurers get to the bottom of many of the Australian desert's secrets, they do not dispel all of its mystery.

Favenc points to the Australian interior, preserving some of its mystery. Leichhardt remains a distant, mystical figure, who suffered 'attacks of feverishness and temporary madness' (after a wound to the hand) and, as he lay dying,

Plate 4.3 Illustration showing the death of Leichhardt, in *The Secret of the Australian Desert* (1896). Favenc's heroes solve the Leichhardt mystery. In this dead, dry, literally and metaphorically lifeless image, the romantic figure of Leichhardt is exposed to the clear light of day, his mystery dispelled, as the mystery of his grave – the Australian interior – is also resolved, at least in part.

'talked a good deal to himself in German' (Favenc 1896a: 95). Even Stuart, the more down-to-earth, British author of the journal they find, had a few mystical experiences in the desert. He recalls a tramp through the desert, 'a dream of stumbling along and helping each other, sometimes talking to the phantoms we all fancied we saw walking with us' (Favenc 1896a: 97). Some of the mystery of the desert remains. Mysteries of lost explorers and cave paintings, like legends of burning mountains and ancient races, offer tentative points of access to the interior. The space they suggest is hazy and alluring, its mystery an invitation to the adventurer and to the colonial geographical imagination.

The geography of *The Secret*, reflecting and accommodating the physical manliness constructed within it, is a physically inhospitable land, replete with physical threats and dangers. An action story, *The Secret* is, in the words of one (particularly complimentary) reviewer, 'brimful of stirring incident and adventure', with 'peculiar attraction' to 'man or boy' (*The Schoolmaster*, cited by Favenc 1896b: ii). Manly endurance and strength are cultivated in a physically rough environment. The trio cross through dense scrub 'covered with the detestable spinifex' (Favenc 1896a: 121). They endure extreme heat and drought on the arid limestone plain, where they suffer but survive thirst and fatigue. They pass through territory so barren that the only thing to eat is human flesh, hence through regions where all the Aborigines are cannibals. They demonstrate that men must be tough and strong, both in character and body, if they are to survive, to avoid either being eaten or degenerating into cannibalism. Morton and Brown prove their strength, for instance, when they shift a large limestone boulder from the mouth of a cave where Charlie is buried, saving his life. The trio defend themselves against physical threats in the desert, a space where 'every step was fraught with danger', and where there are many 'lurking foes'. They survive earth tremors, struggle with 'unknown' beasts and defend themselves against hungry cannibals (Favenc 1896a: 48). Charlie, for example, fends off hungry cannibals, including Chief Columberi, who takes a particular fancy to him. Ostensibly the three white males use violence only in self-defence, but when (they feel) provoked they fight to kill. Invariably, their violence ensures the survival of whites and results in the death of Aborigines, including entire tribes. Violence, a manifestation of the heroes' physical masculinity, becomes an expression of their racial superiority and a vehicle of their colonial conquest.

The geography of *The Secret* is populated with the others of white masculinity, people and places constructed as savage, which the white men either destroy or colonise. Favenc's constructions of savagery and primitivism overlap with those of femininity. As Torgovnick (1990: 156) observes, in her reading of another *fin-de-siècle* adventure story, 'to speak of women in *Heart of Darkness* and to speak of the primitive are, illogically, one and the same thing: fantastic, collective . . . seductive, dangerous, deadly'. Yet, in *The Secret* women are banished almost completely. Except for a few 'gins', the Australian interior is exclusively male space. Women seem to have no part in the story. Metaphorical femininity, which dominates the geography of many adventures, plays a relatively small role in

Favenc's story. Whereas Haggard (in *King Solomon's Mines*) adopted the highly sexualised imagery of men exploring 'breasts' (mountains) and 'nipples' (summits), and entering a 'most wonderful' place, a cave with fertility idols (McClintock 1994), Favenc was sparing with overtly gendered or sexualised language. Favenc's men find a few caves, which they stare into in horror and fascination, and which they are drawn into, much to their peril. But, even in caves, they see savagery rather than femininity, since caves in *The Secret* are the venue for witchcraft and cannibalistic rites. Favenc's neglect of metaphorical femininity reflects his primary interest in material rather than metaphorical geography, while his neglect of women reflects the masculinity of his colonial vision. The tangible, colonial other of white men is the savage Australian land, populated with savage people.

The Australian landscape and its inhabitants are constructed as the others of white masculinity, partly through comparison and conflict between white men and Aborigines. The simple, white masculinity of the heroes is mirrored in the caricatured, black masculinity of the Aborigines. Favenc's savages (as he frequently referred to Aborigines) reflect contemporary constructions of the savage in popular British imperial literature. As anthropologist Brian Street (1975) has argued, popular writers drew upon contemporary scientific and anthropological theories of race, particularly evolutionary theory, which they filtered and modified in their stories.[84] The general result was a stereotyped 'primitive' or savage, inferior to his or her white counterpart, lower down the hierarchies of evolution and civilisation, who was fearful of spirits and mystical beings, and was gullible, cannibalistic, childlike and commonly ugly. Australian Aborigines were generally placed close to the bottom of this racial hierarchy, alongside black Africans. Favenc drew liberally upon such tropes. His setting is marked with the six-toed footprint of the Devil. His Aborigines, variations on the stereotype of 'savage hordes seeking human flesh and dancing gleefully round boiling pots', are the generic ingredients of popular fiction (Street 1975: 76). When the white heroes encounter a tribe of fierce cannibals, the Warlattas, for example, 'an awful feeling of horror came over the whole party' of whites. 'In a wilderness of savage rocks, surrounded by an expanse of desert, almost in the hands of some fifty or sixty fierce cannibals' (Favenc 1896a: 61). These are the 'mixed and degenerate descendents' of a lost civilization – Lemuria, an Australian equivalent of Atlantis (Giles 1988). They have descended into a 'savage' condition, and become 'devil worshippers' (Favenc 1896a: 149).

The white males defend themselves, both psychologically and physically, to avoid being dragged down to the level of, or killed by, the savages, as contrasting masculinities come to blows. The adventurers witness, and play an active role in, the disappearance of the Warlattas, who are trapped in their cave, buried alive with the victims they were about to consume. *The Secret* is set in a space littered with the skeletons and rotting bodies of its original inhabitants. When he sees the remains of a tribe massacred by the Warlattas, Brown reflects, 'What a blessing it is . . . to know that all those wretches who did this are crushed into jelly underneath tons of rock' (Favenc 1896a: 125). Of all the primitivist tropes available to him (Torgovnick 1990), Favenc adopted the most negative, abandoning all traces of the 'noble savage', which survived even in Haggard's Africa. Favenc reused rather than invented savages and whites, neither of which develop into more than caricatures in his story. While the white men begin by defending themselves against savages, they end up by wiping out entire tribes. Genocide, described in *The Secret*, is justified both racially and circumstantially; it is made to seem reasonable under the circumstances. Thus Favenc speeds up the 'survival of the fittest', imaginatively removing stereotyped savages and creating an empty space in which to imagine a new colonial geography.

The setting of a boys' story, Favenc's geography is a space of boyish pleasure, an adventure playground. When the trio of adventurers set off in search of the burning mountain, they do so without serious motives, and are in search of pleasure as much as any mountain. So when they set off, it is 'with light hearts' (Favenc 1896a: 15). Both to heroes, who excitedly contemplate and pursue adventures, and also to readers, regions of adventure are regions of boyish fun. Favenc's setting resembles the playground of a British, or perhaps a colonial Australian boys' school, a space in which masculinities are constructed partly through relationships between boys and men, in the absence of girls and women. School playgrounds and sports fields, like settings of adventure stories, were spaces of male pleasure, while they were also spaces in which masculinities and imperialisms were imaginatively constructed (Mangan 1986). In the Australian geography of Favenc's adventure, as on the sports fields of British public schools, imperialism and male pleasure cannot be separated. Imperial practices, including violence against Aborigines, are designed to entertain and to amuse boy readers. Aborigines, such as the figure on the front cover of *The Secret* (Plate 4.4), illustrate the sensationalism of the story. The figure in the art-nouveau cover design signifies entertainment and excitement, and resembles a Vaudeville performer, a black-and-white minstrel perhaps, with his mixture of black skin and European features; the nose, for example, is more Caucasian than Aboriginal. His posture, like his body markings, suggests passion, witchcraft, evil and sensational mystery. The encounter between white and black is dramatised for the benefit of the white observer. For the mass audience of boys, popular writers emphasised the exotic, exciting and 'savage' in Aborigines at the expense of the everyday, commonplace and therefore dull (Street 1975). Like Haggard, Favenc used blacks as fodder for dramatic, entertaining mutilations and killings, and for comic relief (Katz 1987). The setting of Favenc's narrative, like the characters, is theatrical, an exotic space that accommodates a boys' action story. The coloniser is constructed as white action hero, the colonised spaces and subjects as stage set and *dramatis personae* in an imperial, non-fiction drama.

The masculinity of Favenc's heroes is inscribed in territory the heroes chart, as they make maps. Their masculine, colonial authority is revealed when they physically, but also imaginatively and visually, conquer the desert and its Aboriginal inhabitants. The central image of their imaginative and visual authority is the

Plate 4.4 Front cover of *The Secret of the Australian Desert* (1896). A suggestion of the mystery that is to follow, the cover illustration depicts a naked cannibal with mysterious body markings (red smear and white triangle), holding an unidentified sheet of paper, perhaps a page from the dead explorer's journal.

map they make *en route*. The map, a visual representation from an imaginary bird's-eye perspective, imaginatively controls and possesses the geography of the interior. Representing the land from an imaginary bird's-eye perspective, or from an imaginary prospect point, the map is a gendered image. It naturalises the masculine authority of the colonial map-maker, making him the 'monarch of all he surveys', as Mary Louise Pratt (1992) put it in her study of imperial travel writing. Morton, Brown and Charlie keep a 'rough chart' of their journey, 'compiled every night by dead reckoning' and corrected when possible (Favenc 1896a: 187). The map, an expression and a tool of the men's colonial ambitions – to find gold for themselves and to open up the region, without giving too much away to other prospective colonists – actively constructs rather than passively reflects potential colonial geography. It is selective and deliberately sketchy. The location of the gold reef, for example, remains an unmapped secret. Whereas the explorers in Favenc's *History* 'had a blank sheet to fill up' (Favenc 1888: 7), which they filled with colonial geography, their counterparts in *The Secret* travel in a blank, mysterious landscape. Rather than cramming it with geographical detail, they leave it open, hence accommodating to a range of colonial desires.

In *The Secret*, Favenc constructed an imaginative space in which to map British-Australian colonial geography. The space is simple and uncluttered, a discrete space that is accommodating to a variety of colonial visions. Favenc's settings simplified and denied the ambivalence of colonialism and colonial geography, in which 'real' lines between home and away, men and women, black and white were often blurred (see Dawson 1994). Thus, he constructed a space in which colonial imaginations, like his colonial heroes, could run free. Favenc's heroes are constructed explicitly as colonial figures, who seek and find colonial riches, both for themselves and also for the Australian nation. The trio are cattle-ranching colonists, never the detached 'imperial eyes' that many Victorian adventurers appeared to be (Pratt 1992). They seek and find gold, and keep the location of their gold secret, so they can keep it all themselves. They are openly committed to colonisation in the interior. Morton admits to being an 'optimist and an enthusiast' about Australia; he predicts that 'the end of the coming century will see it settled from east to west throughout' (Favenc 1896a: 213). Brown expects the gold reef he has found will attract the first whites, once they hear about it. He tells Morton that he 'will soon see a road out here and a township too', and reminds him, 'both you and I have seen those things spring up like magic in Australia, before now' (Favenc 1896a: 187). The volcanic landscape of the interior, naturally a space of violent destruction and creation, accommodates Favenc's story of genocide and dream of colonisation perfectly. Favenc's dream in *The Secret*, a dream he reiterated in many of his other works, was that the desert should blossom.

'What a real desert!' said Brown, gazing round on the dreary scene.
'Yes, it's about as hopeless a looking picture as one could find anywhere, at present. And yet, if the artesian water is found to extend throughout

the interior, it will change the whole face of the Australian earth in time. This spinifex would not grow here, but that the climate is so arid that nothing else will grow, and this beastly stuff can thrive without any rain at all. No, burn this scrub off, or clear it somehow, and, with a good supply of artesian water, there are a hundred and one payable products one could grow here.'

(Favenc 1896a: 387)

Favenc has his heroes reflect on how the 'desert theory' has been 'exploded' in Australia. They inform the reader that, for example, 'the date palm will thrive on the shores of these salt lakes so they need not be quite so barren' (Favenc 1896a: 214). In a period when Australia's climate was commonly blamed for producing what one Fellow of the RGS had called the 'feeblest . . . hordes of black, ill-informed, unseemly, naked savages' on the face of the earth,[85] imagining climatic reversal was tantamount to imagining racial renewal. In the arid Australian interior, whites (the surviving members of Leichhardt's party) had merely clung on to their civilised ways, or, worse, degenerated into the condition of cannibals. In the revitalised Australian interior, Australian Aborigines would be succeeded by Anglo-Saxons, who would in turn be racially invigorated. Hence, Favenc's belief in *climate* change is central to his philosophy of producing a new Australian nation, a new Australian race, out of British stock.

In the open-ended, mysterious setting of *The Secret of the Australian Desert*, it was possible to imagine Australia.[86] Like other utopias and dystopias, adventure stories commonly visit the settings of lost civilisations (in *King Solomon's Mines*, for example) and future civilisations (as in *Robinson Crusoe*), and Favenc borrowed from both traditions in *The Secret of the Australian Desert*. While Morton, Brown and Charlie uncover the remnants of a lost civilisation, they are really preoccupied with the future. The interior, violently cleared of its Aboriginal occupants, is a manufactured *carte blanche*, a space in which the explorers are able to dream of a future Australia. In contrast to works such as *The History of Australian Exploration* and *The New Standard Geography of Australasia*, which fill 'blank charts' with explorer stories and explorers' maps, *The Secret* actively creates a blank chart, cultural space in which to imagine Australia. The setting is far from the settlements, and even from existing cattle stations. Not attached to any one colony, it exists between all colonies, and is generically Australian. When the explorers imagine the interior and dream of its future, they imagine and dream of all Australia. Although Favenc insists that the 'Australian desert' could be settled, he presents the desert less as the site of immediate colonisation than as malleable *terra incognita*, a space in which to imagine Australia, to invent the Australian nation.

IMAGINARY AND REAL COLONISATIONS

The geography of adventure does not always remain between the covers of adventure books and magazines. Sometimes it spills over into the 'real' world, into 'real' gendered subjects and spaces, inspiring merchants, investors, travellers, settlers and others to go out physically and become 'empire builders', and also legitimates these colonial acts. Above all, adventure stories such as *The Secret* inspired and legitimated the colonial acts of white men. Like other masculinist adventures, they appealed most directly to the 'masculine, virile and venturesome elements in the population' (Kingsford 1947: 73). They invited 'boyish' men and 'manly' boys, rather than girls and women, and other boys and men, to participate in the fantasies and realities of colonialism.

Although Australians continued to produce and consume stories set in their continental interior throughout the Victorian and Edwardian periods, most Australians lived in towns or settled rural districts. Long after Australia became predominantly urban, its literary images continued to be dominated by bush and outback images of 'droughts, floods and bushfires, goldmines, lost children, Aborigines, squatters and swagmen' (Niall 1988: 547). The imaginative colonial landscape of writers such as Favenc and Macdonald bore only an indirect relationship to most 'real' colonial acts and lives. In practice, adventure stories generally inspired British emigrants to become agricultural or industrial labourers rather than explorers or bushmen, and the stories rarely inspired Australian colonists to take up and head for the forbidding lands in which they were set. The relationship between adventure stories and Australian colonialism was less direct. Adventure stories opened up cultural space in which readers, living in the separate and independent Australian colonies, were able to imagine Australia, as a nation within the British Empire.

Favenc's exploration stories participated in the *fin-de-siècle* nationalism that helped define Australia as a nation, helping to prepare it for Federation in 1901. From the beginnings of British settlement until the last few decades of the nineteenth century, British Australians identified with individual colonies and with the British Empire, but not generally with Australia as a nation (Powell 1988). The Australian 1890s, it has often been said, were years of

> intense artistic and political activity, in which the genius of this young country had a brief and brilliant first flowering. . . . A scattered people, with origins in all corners of the British Islands and in Europe, had a sudden vision of themselves as a nation.
>
> (Palmer 1954: 9)[87]

Nationalists challenged traditional colonial and British imperial loyalties. Nationalist artists and writers, including adventure story writers (Dixon 1995), produced 'national' images, including 'national' landscapes. Living and writing in Sydney, Favenc was actively involved in this Australian national movement. In addition to his geographies, histories and book-length fictions, he wrote stories and sketches for the nationalistic, radical, republican literary and political journal *The Bulletin* (founded in 1880).[88] Consistent with the journal's house style, Favenc's *Bulletin* writing, like much of his other writing, focused on bush and outback images and stories. Like other *Bulletin* writers, and like nationalist poets and painters, he used landscape as a medium in which to articulate ideas of Australia, and emphasised the landscapes of the continental interior. This imaginative geography – reproduced in twentieth-century exploration adventure stories[89] – has proved resilient, as a space in which Australians continue to imagine, and sometimes contest, images of their nation. Characteristically, turn-of-the-century geographer Alexander Macdonald (1907: 5) expressed 'hope . . . in the existence of a wonderful region in the vague mists of the Never Never Land'. And when Griffith Taylor, founder of the Sydney University Geography Department, argued that Australia was not as fertile or receptive to settlement as had been supposed, he spoke in very general terms, about a generic, singular national landscape (Powell 1988).[90]

If adventure stories affect 'real' people and places, it is because they have the power to do so. The power to write is linked to the power to map, to assert certain readings of the landscape, while marginalising other possible readings. The power to map is affected by the place a text assumes among readers, whether they be school pupils or scholars who are formally or informally required to cover canonised texts, or whether they be general readers, attracted to the text for its literary or entertainment value (see Said 1993). Some geographical narratives are denied the power to map, perhaps because they are not canonised, because they sell few copies, or because they are never written or published, but remain in the shadows of history. Paul Carter (1987) illustrates the marginalisation of alternative geographies, the imagined and oral geographies of those without the power to write – including most of the officers, convicts, settlers and others in early colonial Australia. Aborigines are relegated to the final chapter of *The Road to Botany Bay*, to the shadows of white history, apparently as an afterthought. Carter thus parodies the white-Australian historical and geographical narrative in which Aborigines are relegated to the appendix, and highlights the marginalisation of Aboriginal geographies, without presuming to suggest what those geographies might be. Like the famous explorers discussed by Carter, adventure story writers such as Favenc exercised their power to map. Commissioned by the state of New South Wales (to write *The History*) and afforded access to adventure and adventure literature largely by virtue of his sex, race, class and nationality, Favenc was empowered to map and to marginalise alternative maps of the same land.

Favenc's maps of Australia – extreme in their sexism and racism, and in their greedy, violent imperialism – illustrate, or perhaps caricature, Victorian adventure literature. But, as Homi Bhabba (1983) has warned, even the most powerful and seemingly rigid colonial stereotypes may be ambivalent and unstable. In the following chapters, I explore ambivalence and instability in the geography of adventure.

5

AMBIVALENCE IN THE GEOGRAPHY OF ADVENTURE

Home and away in *Daughters of the Dominion*

Although the world of adventure is superficially familiar to readers, who generally find their world views reaffirmed in its bold images and uncomplicated terms, it is not a space in which constructions of identity and geography are wholly stable. For the most part, adventure stories reflect entrenched ways of seeing. Writers such as Ballantyne were content to follow formulae, borrow images – such as stereotypes of Indians and British boys – and respect conventions (A. White 1993). Even the most successful writers of pulp fiction needed to produce several books a year to make a living – usually a hundred or more over the course of a career – and to do this they retold stories and resisted any innovations that might have disrupted sales or taken too long to write. Publishers, too, were content to supply established markets with more of the same, and thereby ensure reliable sales. Partly for these reasons, adventure stories are overwhelmingly conservative. In 1940, George Orwell complained that they were 'censored in the interests of the ruling class' and 'sodden in the worst illusions of 1910' (Orwell 1940: 128). But they are never completely conservative. Masculinities and imperialisms, for example, are not simply reproduced; they are actively constructed and reconstructed, in the geography of adventure. Labels such as masculinist and imperialist, sometimes applied to adventure, are too static to capture this fluidity, in which masculinities and imperialisms are in constant states of flux.

Although superficially confined to male-dominated regions far from home, adventure occupies ambivalent space in which boundaries between home and away, women and men may be fuzzy and unstable. When writers, protagonists and readers of adventure stories observe or transgress these spatial boundaries, they observe or transgress metaphorical boundaries between masculinity and femininity. The polarisation of home and away in adventure literature, particularly British Victorian boys' adventure literature, conforms to contemporary doctrine on private and public space, the spheres of women and men respectively, outlined by John Ruskin in a speech he delivered in 1864. Ruskin argued that 'man's power is active', and that 'his energy [is] for adventure' (Ruskin 1865: 146). Most adventure stories were written by men, for and about boys and men. Girls and women were marginalised. In stories, they are the marginalised sisters, girlfriends, wives and mothers whom the boys leave behind. As readers, they were marginalised, since writers and publishers refused to acknowledge them by writing explicitly to or for them. Critics, led by Ruskin, attacked adventure literature for girls. Girls, Ruskin argued, should be confined not only to the real home, but also to the imaginary home, while their brothers roamed the imaginary geographies of adventure. But while many adventure writers were busy policing the lines between home and away, women and men, others found the geography of adventure more malleable. Transgressing, rather than observing boundaries, they found the geography of adventure an altogether less conservative, confining space. While identities are 'not free-floating', but 'limited by borders and boundaries' (Sarup 1994: 95), boundaries are not immutable. In practice, the sharp distinctions between home and away constructed in boys' adventure stories were artificial and unrealistic, and this was particularly evident to those who had been 'away from home'. Ballantyne, for example, knew the experience of making a 'home' while 'away in the wilderness', and he was sensitive to the recursive relationships between the imperial geography of 'away' and the domestic geography of 'home' (see McClintock 1994; Rees 1988; Van den Abbeele 1980). Although his settings were generally polarised between discrete spheres of home and away, Ballantyne acknowledged in his *Reminiscences on Book Making* that the reality was more complex. He reminisced, for example, that

> Whenever I felt a touch of home-sickness, and at frequent intervals, I got out my sheet of the largest-sized narrow-ruled imperial paper – I think it was called 'imperial' – and entered into spiritual intercourse with 'Home.'
> (Ballantyne 1893: 6)

Ballantyne was aware of the constructed and unstable nature of the boundaries between home and away, women and men. He was aware that adventure settings are not necessarily 'remote from the domestic and probably from the civilised', as Green (1979: 23) put it. While he generally respected and simplified these boundaries in his fiction, others transgressed and shifted them, in order to reshape constructions of gender.

Boundaries between home and away, and between women and men, are particularly fuzzy and unstable in late-Victorian and Edwardian girls' adventure literature, in which constructions of girlhood and womanhood are consciously negotiated. In the last decade of the nineteenth century, girl readers were '[liberated] from the dull fiction that had previously been published' by a new generation of girls' adventure story writers (Major 1991: 32). Heroines of the new girls' adventure literature, like those of contemporary women's travel and exploration literature, appear ambivalent, combining 'masculine' and 'feminine' roles, and transgressing the boundaries between women's and men's spaces, home and away.[91] Heroines seemed like 'women in men's clothing' and they appeared to bring elements of home into the space of adventure. Despite their ambivalence, women's adventures (by or about women or both) are more than difficult-to-classify exceptions to the masculinist norm. They are not mere 'exceptions' that prove the masculinist rule in the literature of adventure, as some critics have suggested (for example, Green 1990: 6). Their ambivalence mirrors and articulates the ambivalence of British colonialism, and the ambivalence of modern adventure literature. Women's travel and adventure narratives illustrate the ways in which adventure stories actively construct rather than passively reproduce gender and imperialism.

Among the new generation of girls' adventure story writers, none was more prolific or popular than the Kentish writer Bessie Marchant (1862–1941), author of at least 150 novels. Marchant's were not the first Victorian girls' adventure stories, if adventure is defined sufficiently broadly to encompass school and other stories with elements of adventure, but they were among the first girls' adventures to be set far from home, as boys' adventures generally were (Jenkinson 1946). Marchant can be compared with British Victorian writers such as Mary Kingsley and Isabella Bird, whose travel and adventure narratives gave girls and women the opportunity to go where few (of their generation) had been before – to far-off lands and seas, in the company of heroines rather than heroes. Marchant can also be compared with the most popular and liveliest adventure story writers of her time. A reviewer in one British newspaper, for example, described her as 'the girls' Henty . . . a writer of genuine tales of adventure with a dash and vigour quite exceptional' (*Daily Chronicle*, cited by Marchant 1923: frontispiece). Marchant's novels, like her many articles and short stories, sold well not only in Britain but around the English-speaking world, making her the most popular British girls' adventure story writer of the time. A huge and enthusiastic readership, composed largely (but not exclusively) of girls and women (Doyle 1968: 190), followed Marchant heroines wherever they went.[92] Marchant heroines went almost everywhere, in tales with such inviting titles as *Courageous Girl: a Story of Uruguay* (1908); *The Girl of the Pampas* (1921); *Norah to the Rescue: a Story of the Philippines* (1919); *Sally Makes Good: a Story of Tasmania* (1920); and *The Ferry House Girls: an Australian Story* (1912), all of these published by Blackie.[93] Unlike her heroines, Marchant never left England, so her geographical knowledge was second-hand, most of it derived from the *Geographical Magazine*, the Bodleian Library and the overseas correspondences she kept (Major 1991).

The settings of Marchant stories were a large part of the stories' appeal, entertaining and amusing readers while addressing more serious matters, particularly those relating to gender and imperialism. Marchant always sought to do more than just entertain, as she once explained to a correspondent: 'To me it is the most thrilling thing in the world to think a plain ordinary woman like myself can sit in this quiet room and talk to girls all round the world, yes, *and influence them*, too' (Major 1991: 33; my emphasis). In their history of girls' fiction, *You're a Brick, Angela!* (1986), Mary Cadogan and Patricia Craig identify Marchant as a radical writer, who empowered girl readers. They write,

> Bessie Marchant . . . specialised in depicting girls who were not the slaves of destiny. Her intrepid teenagers saw the old century out with a flourish,

dashing off in search of adventure to all corners of the globe. Bessie Marchant's 'winds of change' braced girls' fiction for the impact of the Edwardian new woman, who was soon to follow.

(Cadogan and Craig 1986: 57)

Marchant showed that adventures were not necessarily 'boys' own' stories, and that adventure spaces were not exclusively 'boys' own' spaces. Unlike other writers, who polarised the geography of adventure into superficially discrete spheres of home and away, women and men, Marchant mixed home and away, women and men, femininities and masculinities. In so doing, she negotiated constructions of femininity, and girls' and women's identities more generally, in relation to real and imaginary geographies. She broadened the geographical horizons of girls and women, offering them imaginative space in which to nego-tiate their gender, and real space in which to imagine a future, as emigrants and settlers. Marchant's stories illustrate the openness and indeterminacy of the geography and the medium of adventure. Rather than reproducing gendered imperialism, Marchant's adventures reconfigure gendered imperialism.

Marchant addressed gender politics in Britain, challenging constructions of 'women's place' in order to challenge constructions of women's identities, and she addressed gender politics in the British Empire, challenging constructions of 'women's place' in order to challenge the male domination of emigration to British colonies and dominions, notably Canada, the number one destination for contemporary British women emigrants (Jackel 1982). Many of Marchant's adventure stories describe female emigration to, and settlement in, Canada. Marchant wrote stories such as *Sisters of Silver Creek* (1907), which is set on the prairie, *A Daughter of the Ranges* (1906) and *A Mysterious Inheritance* (1914), both set in the mountains of British Columbia, despite British gender roles and conventions that were highly restrictive to women and counterproductive to British-Canadian colonisation.[94] Before turning to read Marchant in any detail, it is necessary to say something about these roles and conventions, particularly as they influenced emigration and settlement in Canada.

WOMEN AND ADVENTURE IN BRITAIN AND CANADA

Victorian girls and women were explicitly confined to geographies of domesticity and enclosure, material space in which their life paths were fixed, and metaphori-cal space in which their gender was fixed.[95] Ruskin, in particular, insisted that girls be confined to the home, both materially and metaphorically. In his view, they should be denied access to books and magazines that might transport them away from the home. Ruskin argued that 'the best romance becomes dangerous' in the hands of girl readers 'if, by its excitement, it renders the ordinary course of life uninteresting, and increases the morbid thirst for useless acquaintance with scenes in which [she] shall never be called upon to act' (Ruskin 1865: 163). His advice to the parent and teacher, then, was to 'keep the modern magazine and novel out of your girl's way' (Ruskin 1865: 165). Ruskin's proclamations on gender and literature were reflected in Victorian adventure books and magazines, in which women were generally confined to homes and excluded from imaginative geography in which they might critically negotiate rigid gender roles, and from material geography in which they might find a measure of independence, for example as emigrants and settlers. Girls' stories set in domestic and quasi-domestic spaces such as homes and schools did accommodate some feminist politics. Late-Victorian and Edwardian girls' school stories, for example, constructed a world of independent women living together, away from the dominating influence of men, where they found some control over their own lives (Auchmuty 1992). But to transgress conventions that confined them to the domestic sphere, it would have been necessary for girl heroes to make a more definite break with home and home-like settings.

For the most part, Victorian girls were excluded from dreams of adventure and ignored by writers and publishers of adventure stories. Their exclusion from the imaginative geography of adventure had nothing to do with their ability to read adventure stories, since girls were more likely to be literate and were more prolific as readers than boys (Vincent 1989; Jenkinson 1946), and it had little to do with their taste for adventure literature. Girls read many of the same adventure books and magazines as their male counterparts, including magazines like *The Boy's Own Paper* and books such as *Robinson Crusoe* and *Tom Brown's School Days* (Reynolds 1990; Salmon 1888; Turner 1957). Indeed, as one 'young lady' explained,

> A great many girls never read so-called 'girls' books' at all; they prefer those presumably written for boys. Girls as a rule don't care for Sunday-school twaddle; they like a good stirring story, with a plot and some incident and adventures – not a collection of texts and sermons and hymns strung together, with a little 'Child's Guide to Knowledge' sort of conversation. . . . When I was younger I always preferred Jules Verne and Ballantyne.
> (Salmon, 1888: 28–29)

Girls' responses to boys' stories were and are, of course, more complex than this quotation suggests. Studies of girls and women reading boys' and men's adventures range, in their findings, from suggesting that they enjoy the exciting male spectacle in much the same ways as their male counterparts (Francke 1995, on recent movies), to speculating that girls and women prefer to have no part in the all-male violence (Savage 1995, on Beowulf).[96] It is impossible to say, and it would be unwise to generalise about, what Victorian girls and women made of the boys' and men's adventure books they read; their exclusion was too complete. For the most part they were neither acknowledged as readers nor represented as heroines of those stories. Girls' literature was shaped less by what girls liked to read than by what parents and educators liked them to read. Likewise, the geography of girls' (as of boys') literature was less a reflection of genuine dreams than of lives that were mapped out for girls. Before the 1890s, the characters of Victorian juvenile literature, and many of the readers, were neatly divided between those who stayed home (or in home-like spaces) – girls – and those who ventured further afield – boys.

Adventure stories offered the world to boys, and both writers and publishers took pride in the imperial careers they inspired boys and men to embark upon. A typical publisher of adventure books and magazines boasted that its publications

> aimed from the first at the encouragement of . . . interest in travel and exploration, and of pride in our empire. It has been said that the boys' papers of the Amalgamated Press have done more to provide recruits for our Navy and Army and to keep the esteem of the sister services than anything else.
>
> (Turner 1957: 115)

Robert Ballantyne's readers, for example, 'were the boys who were to become the soldiers and sailors, the explorers and trail-blazers, the missionaries and bishops, the merchant adventurers, the exploiters, the Word-spreaders, the successes and failures of the great British Empire' (Quayle 1967: 303). Individual biographies, such as that of a man 'whose schoolboy passion for adventure stories eventually led him to Sierra Leone' (Gill 1995: 32), illustrate the point. Editors and correspondents in adventure magazines offered specific advice to boy readers. The editor of *The Boy's Champion Story Paper*, for example, responded to a boy's inquiry about emigration, with the following advice:

> I do know a gentleman who is quite an authority on Canadian farming, and who knows a great many farmers out there with whom he can place respectable lads, but it is only on condition that these lads pay their own passage-money and provide their own outfits. This altogether costs about nine or ten pounds, and any boy with this sum, who wishes to emigrate to Canada, should write to Mr Weeks, and providing he is suitable, this gentleman will arrange for him to go out to Canada with a party of boys, and he will see him safe to his destination. . . . Mr Weeks is keenly interested in the question of emigration, and, being a patriotic Briton, he does not mind spending a good deal of his time in forwarding this excellent object.
>
> (*The Boy's Champion*, 21 February 1903: 617)

Directly and indirectly, then, writers and editors of adventure books and magazines presented boys and men, rather than girls and women, with imaginative geography in which to travel, hunt, fight and, more realistically for most, emigrate and settle.[97]

Adventure literature did some damage to the cause of emigration and settlement in western Canada (as it did in other British colonies), since it mapped anachronistic geography, accommodating to boys and their toys but not to independent women and farmers. Ballantyne and his imitators continued to set stories in a fur-trading wilderness – a region far from home, for boys and men

– throughout the second half of the nineteenth century, even though western Canada was rapidly being transformed into an agricultural frontier.[98] Its vast prairie broken and enclosed, the region was populated with practical settlers, served by a transcontinental railroad, and littered with bones, reminders of the huge herds of buffalo that had disappeared so abruptly.[99] The imaginative transformation of western Canada from fur-trade wilderness to agricultural country might be dated to 1858, when an explorer sent by the British government declared the existence of a 'fertile belt' – an imaginative region that grew more fertile and more extensive as the century progressed, and as Canadian expansionists became more optimistic.[100] As settlers flooded in from Britain and North America, in particular, between 1870 and the end of the century, the western prairies changed dramatically, although the settings of adventure literature remained largely unchanged. Many writers made marginal concessions to the settlement cause, suggesting in prefaces and footnotes that settings had since been civilised, and that true wilderness lay only in the past or in remote corners of the northwest, although they continued to emphasise wilderness settings and remain silent about the farms and especially the towns of western Canada.[101] In so doing, they reinscribed the image of western Canada as a space, far from home and from women, which was inaccessible to female emigrants. This at a time when boosters of Canadian immigration, and women's emigration activists, were acutely aware of problems caused by the male domination of emigration and settlement.

Like its imaginative counterpart in adventure literature, the colonial, western Canadian frontier was dominated by white men. Many men moved west, some with their wives and children, many others alone. Women, on the other hand, did not generally emigrate and settle independently. As a result, western Canada became a male-dominated settlement frontier. Canadian observers perceived a 'shortage' of women in western Canada throughout the pre-war period. This was partly due to lower general emigration rates among women: 50 per cent more men than women emigrated from Britain in the period 1850–1914. It was also due to women's reluctance to emigrate to newly settled parts of the Canadian west. Despite the efforts of many women's emigration societies, which encouraged and assisted women's emigration to the Canadian west, white men outnumbered white women by ratios of up to twenty to one in many parts of the prairies and mountains (Jackel 1982).[102] The male domination of colonialism in western Canada was represented, even exaggerated, in many adventure stories that were published in the Victorian and Edwardian periods. Most writers, striving 'for the Ballantyne style' (Egoff 1992: catalogue entry #112), were content to rehearse ideals of muscular Christianity and celebrate the rule of the London-based fur-trading monopoly, the HBC.[103]

Since adventure stories traditionally mapped western Canada as an implausible destination for women (and sedentary farming people), and since adventure exerted such influence over popular geographical imaginations, promoters of women's emigration sometimes attempted to unmap the masculinist geography of adventure. Promoting women's emigration meant taking on the conventions that kept girls and women home. Ruskin, whose moral views on subjects from aesthetics to gender were by no means unchallenged in *fin-de-siècle* Britain, was criticised by some imperial activists. Salmon, who combined interest in literature with enthusiastic commitment to empire, saw Ruskin's dicta on the place of women as an impediment to empire. The lack of 'Go' in girls' literature, he argued, encouraged a lack of 'Go' – 'a monosyllable signifying startling situations and unflagging movement' – among British girls and women (Salmon 1886b: 515). Salmon's objectives were shared by the Dominion of Canada and the Canadian Pacific Railway (CPR), which regarded women as suitable and necessary immigrants (Friesen 1984; Hammerton 1979; Jackel 1982). They were also shared by some feminists, who worked through emigration societies to promote women's emigration to Canada, which they saw as a source of independence for women. Through propaganda and personal communications (Artibise 1978; Rees 1988), and also through other forms of writing including fiction, emigration activists and feminists attempted to liberate female emigrants. Some, accustomed to the exclusively boyish adventures of writers such as Ballantyne, were convinced that adventure's geography must be unmapped.

UNMAPPING MASCULINIST ADVENTURE

Some writers attempted to erase geographies of adventure. They unmapped adventures, imaginatively clearing the wild west they charted, to open up conceptual space for settlement. Boosters of settlement, but also settlers themselves, condemned adventure stories in which there seemed to be no place for settlers, particularly women settlers. Prairie settler Mrs J.C. Horner, for example, roundly condemned the masculinist adventure tradition in a burlesque of adventure entitled 'dark tragedy – a thrilling story of western life', which remains unpublished, but can be read in the Saskatchewan Public Archives, in her 1911 diary.[104] Horner presented the geography of adventure as an all-male world of domestic squalor and unwholesome yarns, which 'fire the youthful imagination with vicious ambition and a desire to emulate'. She insisted that adventure stories were set in territory 'where now the hand of civilized man holds sway'. Nevertheless, she lamented,

> Thrilling and exciting stories of the warring tribes of red men who previous to that had struggled for supremacy are still told and listened to with open mouth and bated breath by the 'tenderfeet' as they gather round the little bar-room stove or spend the evening in the little 'shack', called by courtesy 'the hotel'.

Well-known propagandists such as R.J.C. Stead,[105] marketing executive for the CPR and later the Canadian government, and Alexander Begg,[106] manager of the CPR publicity office in London, used anti-adventurous literature as a means of promoting emigration and settlement. Stead's 'prairie realism', first seen in sketches and stories he wrote in the first decade of the twentieth century but best exemplified in his 1926 novel, *Grain*, was a self-conscious departure from adventure, as he explained in a letter to his publisher.

> I have just completed a novel of western Canadian life. Although a western tale there is nothing 'wild west' about it, because the wild west of literature owes its existence almost entirely to the imagination of certain low-brow novelists. I have simply tried to paint prairie life as true to conditions as my ability will permit, and I think I may claim that thirty years intimate association with the life of the plains at least to some degree fits me for the task.[107]

Western Canadian critics, many of them writing for newspapers committed to boosting the west, welcomed Stead's departure from adventure. In a review of *The Bail Jumper* (1914), for example, *The Albertan* praised Stead's 'attempts to portray truthfully in fiction the life of the inhabitants of a small prairie town'. The Calgary *Albertan* (12 December 1914) concluded that *The Bail Jumper* would 'serve the very patriotic purpose of creating a new and correct impression of the conditions of living in western Canada, among those people who have so long been deluded by the highly imaginative fiction of a different school of writers'.

But while some settlement-oriented writers distanced themselves and western Canada from the adventure tradition, others wrote settlement adventures, replete with images of homes and farms, railroads and towns, which blurred the once-sharp boundaries between home and away. Emigration and settlement activists such as J.M. Oxley,[108] W.H.G. Kingston,[109] Bessie Marchant, Jessie Saxby[110] and her son Argyll Saxby,[111] all tried to redefine the Canadian adventure story in order to represent and promote new forms of colonialism, including tourism and agricultural settlement. While Marchant, Kingston and Jessie Saxby remapped the geography of western Canadian adventure from the perspective of Britain, to promote emigration, Oxley and Argyll Saxby did so in Canada, in order to promote settlement and imaginatively colonise the region. In *The Boy Tramps; or, Across Canada* (1896), for example, Oxley promotes the CPR's 'all-British route to the Orient' (the boys are headed for Shanghai) by train and steamship, he advertises tourism in Canada, particularly in the Rockies, and he seeks to inspire immigration and settlement in the prairies.[112]

In *The Boy Tramps*, the geography of Ballantynesque adventure is unmapped, clearing space in which to imagine a region of agricultural settlement. Oxley's boy heroes are disappointed to find HBC Fort Garry almost totally gone, and one remarks, 'Wouldn't Ballantyne be disgusted if he were to come back and find that they had torn the old place to pieces, just to turn it into building lots!' (Oxley 1896: 180). The two boys find many of the romantic elements of Ballantynesque adventure to have disappeared or to have been overstated in the first place. Buffalo, for example, are as 'sleepy and spiritless' as 'stall-fed cattle', and 'not a bit fierce' (Oxley 1896: 174–175), while the only bear in the story is confined to a cage, and the Indians are either 'shabby' (Oxley 1896: 213) or Christian, in neither case the 'noble red men' of adventure tradition.[113]

Having exorcised the Canadian west of its boyish and wild adventurous past, Oxley remaps it as a space of rich agricultural territory and material progress. When they leave Winnipeg, continuing their journey west, Bruce and Arthur pass through miles of rich agricultural territory, rehearsing the rhetoric (of explorers and propagandists) of the 'fertile belt' as they go.[114] 'North, south, and west of them lay a world of verdure' (Oxley 1896: 182). The verdant country they traverse is farmed, dotted here and there with prosperous-looking farm houses. The two boys see signs of progress and prosperity wherever they go. The front cover of Oxley's book (Plate 5.1) shows the boys walking the line of the CPR – icon of civilisation and vehicle of Canadian Confederation[115] – as they follow the course of British North American empire west. Landscape is dominated by the railroad, which runs into the distance, transforming nature into scenery that is framed and structured by its geometry. Nature, once the dangerous adversary of adventurous men, has become a picture postcard, the possession of a national railway, something for the boys to admire with the seemingly passive gaze of the tourist. On the front cover, the boys' packs are light because they have sent their luggage ahead by train; they appear to be on nothing more strenuous than a long-distance picnic. Oxley is not just updating adventure to reflect the changing geography of the west. He is actively choosing to include – and celebrate – that new geography, to draw attention to 'civilised' places that were ignored in Ballantynesque adventures.

While writers such as Oxley departed from aspects of the Ballantynesque tradition, using the boys' adventure story as a medium in which to remap western Canada in the popular imagination, others such as Bessie Marchant went further, more thoroughly reworking the western Canadian adventure to subvert boundaries between home and away, women and men, potential emigrants and fields of emigration. Marchant shows how heroines were transformed from adventurous emigrants to practical settlers, as they helped transform Canadian settings from spaces of adventure to prosperous settlements. Including girls and women as heroines, and acknowledging them as readers, Marchant used the medium of adventure to reconfigure the boundaries between home and away, women and men.

DAUGHTERS OF THE DOMINION: AMBIVALENT ADVENTURE

In the context of late-Victorian Britain, where imperial acts were generally boys' acts, constructed as masculine, and where adventure stories were generally boys' stories, Marchant heroines may well have been mistaken for trespassers, girls in boys' stories. According to Green, everyone knows that adventure is a boy's story, and that women's and girls' adventures are nothing more than 'exceptions' to this historical rule (Green 1990: 6). Green calls adventure 'the

Plate 5.1 Front cover of *The Boy Tramps* by James Macdonald Oxley, published by Musson (Toronto) in 1896. Bruce and Arthur, their packs light, are on a long-distance picnic.

rite de passage from white boyhood into white manhood' (Green 1991: 41); but Marchant's adventurers pass, instead, from white girlhood to white womanhood. In the context of a gender-divided literature, where girls stayed close to home or school while their brothers ventured further afield, Marchant's heroines may have seemed out of place, ambivalent figures in ambivalent narratives. They were not trespassing, though, because adventure is not intrinsically masculinist. Rather, it is a gendered narrative in which masculinities and femininities, male and female spaces and roles, are constructed and negotiated. Marchant transgresses boundaries through ambivalent images and narratives. She negotiates entrenched constructions of gender and imperialism – girlhood and womanhood, home and empire, gendered imperialism and imperial gender – and she does so entirely within the medium of adventure, in a quintessentially modern, colonial adventure story, which illustrates the malleable and fluid nature of the geography of adventure.

Since there is relatively little variation between Marchant's adventure stories, particularly those set in Canada, which follow tried-and-tested formulae, a single story can be considered representative, and is therefore worth reading in detail. *Daughters of the Dominion; a Story of the Canadian Frontier* (1909) is broadly representative of Marchant's Canadian stories, which vary little in either plot, characters or (in a generic sense) setting. *Daughters of the Dominion* is the story of a young – initially 17-year-old – woman,[116] Nell Hamblyn, who leaves a ranch in the American Rockies, and emigrates to Canada, where she works as a domestic servant, a telegraph operator and, later, a restaurateur. The story incorporates many elements of formula juvenile adventure: buried treasure and riddles of the past, last-minute rescues and hairbreadth escapes, physical violence and attempted crime, chance encounters and mistaken identities, good guys and bad guys, and, finally, marriage and happiness. *Daughters of the Dominion* was published amidst a 'flood of books with girls as the chief protagonists' (Egoff and Saltman 1990: 10), which were written and set in Edwardian Britain and North America. The most prominent of Nell's literary ancestors, and the most famous of all Canadian girls' story heroines, made her debut the year before Marchant's Nell. *Anne of Green Gables* (1908), the 'spirited redhead' created by L.M. Montgomery (1874–1942), 'added a note of girlishness and mischief to Canadian children's books that was hitherto lacking' (Egoff and Saltman 1990: 304). Like Anne, Nell grows up a cheerful orphan dressed in ragged clothes and living in a rough clearing, until one day she finds herself transported to an idealised, rural Canadian setting, where she ultimately passes from girlhood to womanhood, and from insecure adolescence to more secure and happy adulthood. Compared with Anne, however, Nell is more adventurous, bolder in taking her fate into her own hands. She leaves the farm where her grandfather brought her up, and strikes out alone, working as she goes.

Daughters of the Dominion begins, as many adventure stories do, with a geographical fantasy. Like Anne of Green Gables, who 'could never have lived' in the rough clearing of her childhood if she 'hadn't had an imagination'

(Montgomery 1908: 57), Nell lives, survives and ultimately escapes through her imagination. Nell has a few books, one an atlas, which she pores over for hours. 'Geography was so fascinating that she had to look upon it as a play task, the study being altogether too delightful to be regarded as work' (Marchant 1909: 186). Like Robinson Crusoe, whose 'Head began to be fill'd very early with rambling Thoughts' (Defoe 1719a: 2), Nell imagines distant places, and often thinks of leaving home to find them. While the image of a boy staring at a map and dreaming of adventure is a trope in Victorian adventure literature, the image of a girl doing the same is less common and more radical. When boys dreamed of leaving home – rebelling against the confinements of their domestic life and the rule of their father – they were essentially doing what was expected of them: imagining rites of passage from boyhood to manhood. Their rebellion, whether against biological or symbolic fathers, was ultimately conservative, maintaining rather than subverting the patriarchal social order. Girls, on the other hand, were expected to remain within the home and under the authority of the male head-of-household. To stare at maps and dream of adventure was, in effect, to imagine transgressing British gender boundaries.

For women, to imagine emigration was to imagine transgressing rigidly bounded spaces and spheres. Marchant is very conscious of this. One Marchant heroine, particularly explicit in her defiance of gender boundaries, tells her brother – who is cynical about her plans to emigrate to Canada[117] – that he is 'quite early Victorian in [his] ideas of what girls should or should not do' and insists that he 'widen [his] outlook' (Marchant 1917: 13). Clearly the idea of female adventure, even when tied to the patriotic business of empire building, was both feminist and radical. Nell is conscious of the conventions she must transgress if she is to leave. Since she is a girl, and since girls are taught that their place is in the home, the thought of running away makes her shudder. Nell asks a passing stranger who stops for rest and food, 'Where should I run to? And who would take me in? A girl isn't able to shift for herself and defy the world like a man' (Marchant 1909: 24). To Nell, the idea of striking out into the wilderness seems like something out of a boys' adventure story, something only a boy could do.

In her geographical fantasy, her dream of adventure, Nell transgresses boundaries, as adventurers have always done; in this respect, she is a typical adventurer. Adventure, in the words of Zweig, is 'an anarchic dream of heroic energies and escape' (Zweig 1974: 14). Nell, like adventure heroes and heroines before her, entertains a dream that is essentially anarchic, and she demonstrates the possibility of escape. Her eventual departure is forced by circumstance, although it remains a radical move, and she knows it. Nell does not abandon her domestic responsibilities, unjust and unhappy though they are, and she does not rebel against her (grand)father – Doss Umpey – as did male heroes from Daniel Defoe's Robinson Crusoe to Robert Ballantyne's Charley Kennedy (in *The Young Fur Traders* 1856). But, when Doss Umpey disappears (under suspicious circumstances) and an unpleasant family of settlers occupies the farm, Nell feels she has no choice but to leave. The settlers want her to stay, to become their domestic drudge, but she takes the initiative to leave. Thus Nell is able to plan her departure without feeling she has abandoned genuine domestic responsibilities or a genuine father – since Doss Umpey has already left and since, it later transpires, he is not her 'real' (biological) (grand)father at all. Thus Marchant arranges Nell's departure so as to minimise the threat it poses to the gendered social order. Nell leaves only when circumstances conspire to force her out of the home. Nevertheless, her final departure is an assertive statement, bolder than the migrations of predecessors such as Anne of Green Gables, who moved to Prince Edward Island as the passenger of a man who had chosen her, rather than vice versa. Having made her bold decision, and 'quite prepared for her plunge into the Unknown', Nell sets out with a few clothes and some bread she has baked.

Nell transgresses gender boundaries in her travel and work. Her first day on the road is a difficult one, and it tests her powers of endurance.

> Nell was very tired. Since early morning she had tramped steadily, pursuing that apparently unending trail. Sometimes the way had been up steep ascents, over high ridges, where big boulders stuck up among the trees; then it would drop to lower ground, and skirt wide swamps.
>
> (Marchant 1909: 58)

Later, confined by domestic and telegraph work, Nell often longs to be out in the fresh air. When a railway official needs to inspect the telegraph two miles from the depot, Nell volunteers to accompany him as 'the prospect of a few miles' run on snow-shoes was alluring to her, after her long days of imprisonment in the warm, stuffy little office' (Marchant 1909: 137). In the spring, Nell cannot wait to get outside. 'But for the duty which chained her fast to the little office at the depot for twelve hours out of every twenty-four, she would have been out-of-doors the whole day long' (Marchant 1909: 145). Leaving home, striking out and eventually finding herself in the remote outposts of Bratley Junction and Camp's Gulch, Nell enters what – in society and in literature – was conventionally regarded as male territory.[118] She also ventures into what seem to be male shoes – conventionally male roles.

Despite her occasionally Amazonian appearance, Nell is not a girl in boy's shoes, but a girl assuming the assertive and strong role demanded by the adventure story in which she finds herself; again, she is a typical adventurer, despite her sex. Writing girls' adventures, Marchant did not appropriate a boys' story. Neither the female adventure story nor the strong heroine were new developments in the 1890s; they represented the revival of a tradition that was all but extinguished by the masculinisation of adventure literature in the Victorian period. Variously Amazonian women were characteristic of female adventure, as it was traditionally written. In women's military adventure stories, for example, women commonly dressed as men (Wheelwright 1987, 1989). In other adventure stories such as Charles Dibdin's eighteenth-century Robinsonade, *Hanna*

Hewit; or, The Female Crusoe (1790), female narrators claimed to have 'male minds' and 'male styles' (of writing) which, they suggested, qualified them as the narrators of adventure.[119] In broad historical perspective, although not in the Victorian juvenile literature, Nell's Amazonian appearance is typical rather than exceptional, and she has many counterparts among the heroines of girls' and women's adventure literature. Although she sometimes looks and behaves like a boy or man, Nell is not a variation on a male theme.

Nell is a strong, womanly – not boyish or manly – figure. Initially a teenage girl with a will to make something of herself, later a mature woman, her adventure is a rite of passage and a test of strength. As a child, Nell is rugged, very much like Montgomery's Anne, who is an 'odd little figure in the stiff, ugly dress, with the long braids of red hair and the eager, luminous eyes' (Montgomery 1908: 33). Anne is literally the next best thing to a boy, since the colonists who take her in want a boy but cannot get one. Nell, too, has many characteristics more commonly associated with boys. She is handsome rather than pretty, with plain looks and 'thin muscular arms' (Marchant 1909: 21), and she has a 'sweet, low voice' and 'luminous eyes' that make up for 'the defects of her face and figure' (Marchant 1909: 9). Nell never aspires towards 'ladylike' dress or behaviour. She has, for example, no reservations about helping the railway guard unload boxes from the train, behaving in a manner that another woman (Miss Simpson, the telegraph operator from New Westminster) condemns as 'fearfully unladylike' (Marchant 1909: 117). She compares favourably with effeminate characters, including a 'soft' office boy who is afraid of wearing snow-shoes, is reluctant to exert himself, and is described as 'a sickly-looking youth', too 'delicate' to cope with the rugged Canadian west (Marchant 1909: 117, 171). Through her adventure, Nell is transformed from a rough-looking American farm girl to a 'well-dressed, eager-faced', married Canadian woman (Marchant 1909: 166). Her 'womanliness' is manifest in displays of 'womanly tenderness' (towards patients), 'womanly' strength (to face the death of Pip the dog), and a desire to 'excel in all the womanly occupations', such as cooking and needlecraft (Marchant 1909: 21, 147). She is shown on the front cover (Plate 5.2) as a 'womanly' figure, standing boldly with nature underfoot, axe by her side, in a rugged (but not threatening) Canadian landscape. An assertive and strong figure, Nell is reminiscent of boyish and manly, male adventure heroes. However, her womanliness is not a copy of manliness and it is not the same as manliness. Both womanliness and manliness are variants of the Victorian, gender-neutral concept of humanliness, or, more specifically, Christian humanliness. Humanliness, 'counter not so much to womanliness as to effeminacy' (Vance 1985: 8), encompassed that which was 'best' and most 'vigorous' in humans (Haley 1978; Vance 1985: 1). It was generally regarded as 'a good quality', with 'connotations of physical and moral courage and strength and vigorous maturity' (Vance 1985: 8). In the strong, vigorous image of Nell, as in many of her other heroines, Marchant reclaimed an ideology and a literary tradition that had been appropriated and masculinised in the Victorian period by boys and men.

Plate 5.2 Front cover of *Daughters of the Dominion*, this edition published by Blackie (London) 1909. Nell, in thick skirts with axe at her side and nature underfoot, is shown to be 'womanly', not 'ladylike'.

Nell never quite abandons herself to adventure, as some of her male counterparts in adventure fiction tend to do. She does brush with adventure, though. Nell's big adventure takes place at Camp's Gulch, a remote telegraph office where she is posted. One day she finds herself responsible for some valuable goods, which are being stored in the railway shed. That day, two men arrive with a coffin, said to contain a 'Dead Chinaman', to be stored overnight in the same shed. Nell's suspicions are aroused when she notices air holes in the coffin. She binds the coffin in heavy chains, and prepares to telegraph a warning to the railway company. When she finds the telegraph wires have been cut, she is not deterred. She sets off down the line, finding the point at which the wires were sabotaged, and proceeds to send a message through the open wires. This message is received by Nell's friend, Gertrude Lorimer, the telegraph clerk at Bratley Junction, fifteen miles away. When, eventually, two of the men at Bratley Junction manage to get a locomotive up to Camp's Gulch, they find Nell lying unconscious on the tracks, with injuries to the face and arm. Nell has saved the valuable goods, and captured the would-be thief, hidden in the coffin, who turns out to be a man wanted for fraud by the local mining company. In the process she has become a heroine. 'They were saying the kindliest things about her, magnifying her into a heroine, while she [was] lying there with her broken wrist, hurt jaw, and torn ear' (Marchant 1909: 233). At this point, however, Nell's adventures come to an abrupt end. She loses her job as a telegraph operator because she cannot hear clearly, she gives up ambitions for an education and a professional career, and resumes her domestic labour.

Although Nell never quite abandons herself to adventure, this does not make *Daughters of the Dominion* a watered-down tale that might be regarded as a compromise in deference to the girls' literature market. Rather, Nell's is a colonial, modern adventure story. Nell does retain many of her domestic ways – practical clothes and a strict housework ethic – even while she is on the road and in the rugged mountains of British Columbia, and her first adventures are purely domestic. Soon after leaving home, she arrives by chance at the home of a sick woman, just in time to clean the house and save the woman's life, both of which the man of the house seems incapable of doing. 'You came in the very nick of time', the doctor tells her, as he accompanies Nell to another farm where her help is also sorely needed (Marchant 1909: 72). Nell delights in domestic work. 'Washing, baking, sweeping, scrubbing, the days passed like a dream to Nell, and she was happier than she had been in all the years since her father died' (Marchant 1909: 101). Later, she works as a 'girl telegraph clerk' at Bratley Junction, a railway depot in British Columbia's 'rugged' mining country near Lytton (Marchant 1909: 49). The geography of Nell's adventure, replete with farms and houses, offices and railroads, contrasts with that of many Canadian boys' adventure stories, which are set in undiluted wilderness, populated by 'wild beasts and wild men' (Ballantyne 1856: 24). Marchant's tale is set on the *edge* of the British Columbian wilderness – 'the extreme end of civilisation' – far from the 'luxuries of civilisation' found in towns such as New Westminster, but not actually *in* the wilderness (Marchant 1909: 167, 49). There are no hostile Indians, no wolves and no bears, not even the relatively harmless, black variety. In the winter there is no problem with cold, just a little snow that gives Nell a chance to demonstrate her mastery of snow-shoes. Nell's adversaries are generally moral rather than physical. She encounters laziness, dishonesty, lawlessness and cruelty, mainly in Doss Umpey and his friends. Perhaps, in the vaguely phallic image of the stick that wounds her, she is exposed to sexual danger, although such danger is never made explicit. But Nell's response, even to stick-wielding male adversaries, is gentle and compassionate. When Doss Umpey returns to wound Nell, her response is to forgive him and, when he shows signs of dying, to nurse him through his last days. Nell shows herself to be a domestic heroine. Much of Nell's time is spent indoors: at work, at home with her landlady (talking, eating and sleeping) and at the houses of nearby settlers, whom Nell helps out with child-minding, cooking, cleaning and mending. Nell's 'adventure' begins to appear quite tame, while its heroine appears ambivalent, even reluctant in her role. She never quite leaves home, never quite abandons so-called civilisation. In one illustration (Plate 5.3), for example, she is depicted inside a hut, surrounded by the paraphernalia of domesticity, even as her adventure is taking place. The geographical and metaphorical polarisation between home and away, which structures many adventure stories, is disrupted in *Daughters of the Dominion*. But the same can be said of other modern adventure stories, such as *Robinson Crusoe*, in which the quintessentially modern, colonial adventure hero spends much of his time engaged in practical labour, much of it around the home.

The contradictions and tensions in Nell's identity and in her story mirror the contradictions of colonial adventure – emigration and settlement – and of modern adventure more generally. Female emigrants, even more than their more mobile, male counterparts, knew the contradictions involved in dramatically leaving one home, only to settle down and build a new home elsewhere. However sedentary they may have been before their journey and after it, emigrants had to be at least a little bit adventurous for a time, setting out into the unknown, leaving their society and their civilisation behind them. They then had to settle down, to abandon their adventurous ways, for a quiet, practical, domestic life. As one emigration activist put it in 1883, 'To leave their homes and come out here is,' for British farmers, 'like performing *Robinson Crusoe*'.[120] Indeed, Nell's ambivalent story closely parallels that of the archetypal modern adventurer, Robinson Crusoe. *Robinson Crusoe*, like *Daughters of the Dominion*, combines adventure and survival, emigration and settlement. At times, it is a story of mere survival, its main character a reluctant hero, who avoids rather than seeks adventure. Indeed, critics have debated whether *Robinson Crusoe* was an adventure story or a survival story (Zweig 1974; Green 1979). Although Nell is just as happy with a mop and a bucket as she is on the trail of adventure, this is a quality she shares with the original Crusoe. Like Nell, Defoe's hero sets out a restless young man – an emigrant – and becomes a practical, mature Christian – a settler.

Plate 5.3 Illustration in *Daughters of the Dominion* (1909), with the caption 'I heard a little about a friend of yours'. Nell, perched on the *edge* of the British Columbian wilderness, is surrounded by the paraphernalia of domesticity.

Nell is politically ambivalent, sometimes transgressing contemporary conventions of gender, at other times observing them; she is far from radically feminist. An orphan, forced out of her American farm, her criticism of British constructions of gender and gendered space is as cautious as it could possibly be. The orphan, a popular literary convention, offered women writers such as Marchant 'physical and psychological room for manoeuvre' (Bolt 1995: 25), freeing them to invent 'new women' (Reynolds and Humble 1995), but excusing them from acknowledging or confronting the domestic bonds and gender conventions that held many single women in place. Nell's adventures are calculated to transplant her from one domestic sphere to another, while causing minimal disruption to the British-Canadian social order. Nell's ambitions, which go beyond the domestic, are permitted only to the extent that Nell is motivated to find her way to British Columbia. Through her adventures Nell discovers some of the possibilities open to contemporary women, but also some of the limitations. Wounded by the villains, Nell decides that she has over-reached herself. Her recovery is slow, since 'the shock and strain of her adventure at Camp's Gulch proves too much for even her intrepid spirit' (Marchant 1909: 251). She is depicted (Plate 5.4) after her brush with villains as the clichéd image of a woman lying helpless on a railway track, viewed from the perspective of the men who save her.

What follows is a period of retrenchment, as Nell renounces some of her adventurous spirit. She goes through a long period of convalescence, as her health slowly returns. Nell's poor hearing is always a reminder that 'she had rather seriously overdone her strength that day' (Marchant 1909: 348). Unable to continue working as a telegraph clerk, but with a reward from the mining company (for capturing the fraud), and compensation from the railway company (for her partial loss of hearing), Nell finds herself with enough money to begin an education. In her heart, she has always harboured an ambition to become educated, to leave the domestic world behind, and perhaps to enter into some (male-dominated) profession. But, when she consults him for advice, the local doctor advises Nell against pursuing medicine as a career. He tells her, 'there is always a crying need for bright capable women in what are mistakenly called the humbler walks of life' (Marchant 1909: 257). Nell is disappointed, but 'she had the common sense to know how truly the doctor had spoken' (Marchant 1909: 258). Nell 'remembered her father's words about seeking Heavenly guidance in the grave decisions of life' (Nell's father had been a Christian minister) (Marchant 1909: 259). She turns away from adventure and accepts not only the guidance of the doctor, but also that of God, both of whom advise Nell to pursue practical work, specifically domestic work. Nell devotes herself to all forms of domestic work, and like Robinson Crusoe she gains practical Christian knowledge through her labour. She adopts the remaining members of the Lorimer family, recently orphaned, and invests her reward and compensation money in a business, a restaurant for the miners. The restaurant business is to be Nell's niche in that mostly male mining district of British

Plate 5.4 Nell's injury, illustrated in *Daughters of the Dominion* (1909). Nell presents the clichéd image of a helpless woman, viewed from the perspective of men who come to her rescue.

Columbia. Nell proves herself as a successful businesswoman and domestic heroine combined. Eventually Dick Bronsen, whom Nell encountered in the opening scene of *Daughters of the Dominion*, asks for her hand in marriage. 'And so they were betrothed' (Marchant 1909: 352). Thus Nell resumes her place in society, and resolves to respect its conventions.

Nell's story was intended to inspire girls and women to emigrate and settle. Her story shows female readers that they, like their male counterparts, can be adventurers, and they, too, can imagine emigrating. Nell is a role model for girls to follow, and an image of Canada's preferred immigrant, as defined by the CPR and other boosters of pre-war western Canada. Brought up in virtual isolation on the American side of the western 'frontier', she is practical, accustomed to hard work and rural living (which she prefers to city life), and is keen to make something of herself. While other Marchant heroines emigrate to Canada from Britain and eastern Canada, all come from rural environments, and share Nell's outdoor ruggedness. Nell is particularly qualified – by virtue of the gender roles and gendered spaces she defers to, and also those she transgresses – to perform the balancing act between home and away, domesticity and adventure, emigration and settlement. And she is receptive to the idea of Canada. Marchant has Doss Umpey advise Nell that 'Canada is a land of promise for young people. Then, English law, by which, of course, I mean Canadian law, is kinder to lone women and girls than American' (Marchant 1909: 13). Nell learns of Canadian opportunities, available to hard-working people with the will to work, from the doctor she meets in America. He reassures Nell that she will have no trouble in getting work when she crosses the border. Canada, as Nell finds it, is a land free of environmental hazards such as bears and snowstorms, and a land with much to offer the immigrant, including resource-rich 'fertile lowlands' and 'very rich' copper ore deposits (Marchant 1909: 167). In case she was not making herself perfectly clear, Marchant prefaced the tale with the reminder that Canada offers a home to all, whether they be male or female.

> CANADA is a great mother; there is room in her heart not merely for her own children, but for the needy of every nation. They may all come to her and find a home, if only they will work to earn it.
>
> (Marchant 1909: 4)

'No longer content to be simply backers-up of male empire builders', it has been suggested, 'girls were seeking new worlds of their own to conquer' (Cadogan and Craig 1986: 59) – worlds mapped by writers such as Bessie Marchant.

CONCLUSION: NEGOTIATING GENDER, CULTURE AND IMPERIALISM

Rather than condemn the geography of Canadian adventure as masculinist and imperialist, excluding women and settlers, Marchant moves within its open-ended terrain, unmapping and remapping constructions of gender identities and gendered imperialism. Marchant heroines redefined some (but not all) of the gendered terms on which British patriarchy and British colonies were founded. Blurring and shifting the boundaries between home and away, Marchant subverted the polarisation of men and women into separate spatial and economic spheres. Blurring and shifting the boundaries between women and men, Marchant subverted the polarisation of gender identities, which, it has been argued (Wheelwright 1987), were increasingly being seen as biologically rather than culturally determined, hence immutable. She used boundaries between home and away, women and men, as dynamic cultural spaces in which to map, unmap and remap gender identities and gendered spaces.

In Canada, Marchant opposed a masculinist tradition of adventure that had made a strong impression on popular geographical imaginations. She may have changed the way girls and women thought about western Canada, although she did not suddenly unmap the literary landscape of boyish adventure. The western Canadian literary landscape continued to be dominated by Ballantynesque boys and men, and by the iconography of masculinity – by horses rather than houses, as Canadian literary critic and writer, Robert Kroetsch has put it.[121] This gendered colonial space is offered, materially and metaphorically, to boys and men, and denied to girls and women. Marchant, writing for girls and women, attempted to unmap and remap western Canada in the geographical imaginations of girls and women. Her appeal to girls and women, rather than all readers, is a reminder that no literary mapping is definitive, that different groups have different ways of seeing, and that even the dominant ways of seeing are countered with currents of resistance (Francis 1989; Silver 1969). Thus, any mapping, unmapping or remapping is partial, at best affecting some geographical imaginations some of the time.

Stories such as *Daughters of the Dominion*, unambiguous though they often seem, have no singular meanings, so they allow no definitive readings. Marchant wrote largely for Victorian and Edwardian girls and women, and I have interpreted her books with those readers in mind. But, just as girls and women read so-called boys' stories, boys and men read Marchant's so-called girls' stories. A copy of *Daughters of the Dominion* that I came across in the University of British Columbia Special Collections, for example, is addressed 'To Duncan, From Harold, Wishing him many happy returns of his 19th birthday, 1912'. My reading of *Daughters of the Dominion*, and Marchant's preferred meaning, may not be the same as Duncan's. Instead of being inspired to throw off the shackles of domesticity, this 19-year-old reader may, for example, have been disappointed to receive a 'girls' story' for his birthday; alternatively, he may have enjoyed the novel spectacle of heroines roaming the wilderness; he may have found Marchant's heroine sexually exciting; he may have been influenced to support Marchant's feminist project; he may have cross-identified with the heroine in some way; he may just have been entertained and amused. But, whatever his response, it was unlikely to have been that of the girl reader assumed by Marchant. Since adventure stories have no single, fixed meanings, readings of adventure should not end with interpretations of dominant meanings and authorial intentions.

6

READING AND RESISTANCE
Anarchy and anti-imperalism in French
extraordinary voyages

Adventure can be read as a radical narrative, the geography of adventure as a space of anarchy and a site of resistance. Both in its general imagery and also in its particular images, adventure is peculiarly adaptable to articulating critical politics. Adventure's realistic images are capable of articulating critical ideas that cannot be expressed in more abstract terms, whether because they are not precisely formulated, because they might be censored, or because they would be less readable, therefore less effective in that form. In their perilous journeys, adventurers depart violently from the world they know, but find something on the other side, survive the ordeal and return to tell the tale. In general, they transgress the margins of the human world and the 'confinements of the human situation' (Zweig 1974: 16). Specifically, they transgress particular boundaries and conventions. For example, a given adventure story may be generally anarchist on one level, or specifically opposed to slavery in British colonies, on another, or both.

I argue that adventure stories *can be read* as critical narratives and that geographies of adventure *can be read as* spaces of resistance, not that adventures *are* – sometimes or always – intrinsically critical. Adventure stories, like other texts, are consumed in particular contexts, in which readers, critics, librarians, publishers and others actively read and interpret them. Thus, adventure stories become meaningful in historical and geographical context, among situated textual communities (see Stock 1986). The meanings of *Robinson Crusoe*, for example, are historically and geographically constructed. *Robinson Crusoe* does not have the same meaning in Britain today as it had a century ago, or two centuries ago, as I explain in Chapters 2 and 7. Reading *Robinson Crusoe* in context, however, I emphasised that story's dominant meaning, and thus privileged the dominant textual community, at the expense of alternative readings and textual communities. Effectively, I deferred to the hegemonic reading of *Robinson Crusoe*. The hegemony of an historically situated text is derived not only from an ability to block alternative narratives – a power Said draws attention to in *Culture and Imperialism* – but also from an ability to dictate its meaning, power to block alternative readings. But power over the meaning of a text is never absolute; alternative readings are never completely blocked. As Michel Foucault

(1978: 96) has argued, 'Resistances . . . are the odd term in relations of power; they are inscribed in the latter as an irreducible opposite.'

Dominant meanings of adventure are resisted by writers who retell stories with the intention of subverting them (although their subversions may be lost on readers), but also by readers. In Chapter 7, I show how writers have appropriated adventure narratives, in projects of post-colonial resistance, beginning within the colonial discourse itself. In this chapter, however, I stay with readers, and explore ways in which adventure stories can be read as critical narratives. I pay most attention to readers who wrote critical reviews and commented on adventure stories in public, in print, partly because these were the readers who left relatively clear traces of their acts of reading (most readers left no traces at all), partly because these were the powerful opinion leaders whose readings defined what books were about and dictated their dominant meanings. Histories of individuals' reading may take the form of oral histories (Lyons and Taksa 1992), reader-response psychologies (Bratton 1981) and autobiographies, both critical (Jackson 1990) and psychoanalytic (Dawson 1994), although these are most successful when dealing with periods for which readers are still alive, since they all attempt to somehow get inside readers' minds. It is difficult, often impossible, to get close to the minds of readers who are long since dead. Evidence of reading, reported by critics and reviewers of children's literature such as Edward Salmon, is as much about the critic's readings and judgements, as about the children's. Readings by critics such as Salmon are important, however, because the dominant reading of a text is not defined by the sum of its individual readings. While individual readers may be creative, active agents who participate in the construction of textual meaning, they also belong to textual communities, and generally share interpretations. It is the shared meanings of textual communities – the readings they construct and the readings they effectively marginalise – that I emphasise in this chapter.

Resistance to the hegemonic identities and geographies mapped by powerful adventures, whether in the form of critical readings or critical writings, is the 'irreducible opposite' of – not a chronological response to – adventure stories, and the cultural spaces they construct. It has often been argued that adventure stories were traditionally conservative, but were subsequently appropriated by more critical writers. Paul Zweig (1974) laments what he sees as the watering down of adventure in some quarters, in the modern period. Martin Green (1990) traces resistance to modern adventure in the form of twentieth-century 'anti-adventures'. Other critics trace the emergence of a critical tradition in imperial adventure, both in the genre as a whole (A. White 1993), and in the literary careers of individual writers including Jules Verne (Butcher 1990), beginning in the 1870s. Andrea White interprets some late-Victorian and Edwardian British adventures as anti-imperial. She shows how Rider Haggard and especially Joseph Conrad used the medium of adventure to deconstruct imperialisms and imperial subjects which had previously been constructed in adventure stories. She argues that adventure evolved from a conservative to a critical narrative in the second half of the nineteenth century, from reproducing to subverting the imperial *status quo*. Prior to Haggard, adventure was 'the genre that had always defended the status quo' (A. White 1993: 99). White argues that Haggard, particularly through graphic attention to imperial violence and open admiration for African warriors, used adventure to express an ambivalence towards – not an unqualified endorsement of – British imperialism. Conrad, she argues, used the medium of adventure roundly to condemn European imperialism in Africa.

White's history of adventure demonstrates the possibility of reading particular adventures as critical stories, although, the corollary of this, it also suggests the possibility of other, more conservative readings of the same stories and the same spaces. The critical component of stories by Haggard and Conrad, however overt it may seem to the late twentieth-century reader, lies not (only, or at all) in the intrinsic properties of the text, nor the intentions of the author, but in White's reading. Different readers have constructed Conrad and Haggard differently. Conrad's adventures appear to many late twentieth-century readers as damning indictments of all European imperialism, but when they were published they were commonly seen differently, for example as a narrow criticism of Belgian – not British – imperialism (Torgovnick 1990). And the critical component in Haggard, while evident to White, is less apparent to most contemporary readers, who continue to regard works such as *King Solomon's Mines* as racist and imperialist. White does not uncover the 'true' Haggard; she presents one possible Haggard. Her interpretation of Haggard's violence, for example, is cogent but never definitive (no reading could be): the violence could always mean something else (see, for example, Katz 1987). So, while seeking to show how particular texts are either critical or conservative or somewhere in between, White demonstrates the agency of readers to make a text critical.

No adventure story is intrinsically either conservative or critical, since no adventure story has a singular meaning. My readings of *Robinson Crusoe*, *The Young Fur Traders*, *The Secret of the Australian Desert* and *Daughters of the Dominion* could all be countered with other, alternative readings. While White reads Ballantyne as uncritically pro-imperial, reproducing ideologies of imperial subjectivity, and while I read Ballantyne as asserting an uncritical ideology of masculinity, it is possible to read the same Ballantyne tales differently. For example, Ballantyne's manly boys need not be read as figurative images of Christian manliness – the author's stated intention (Ballantyne 1893). For readers who misunderstand or disregard the intended or 'official' meaning, heroes such as Charley Kennedy and Ralph Rover may, for example, become objects of desire, aestheticised and sexualised. Particularly since the *fin-de-siècle* 'homosocial schism' in Britain, in which male–male relationships were increasingly polarised and policed, and 'interest in men' tainted with the new concept of homosexuality (Sedgwick 1985), heroes of early-Victorian adventures such as *The Young Fur Traders* have lost their innocence, at least for some readers. To homosexual critics such as J.A. Symonds and the Reverend E.C. Lefroy at the turn of the century, and probably to many earlier readers, British Victorian adventure heroes were objects of desire, unmistakably homoerotic. To Jean Genet, in the twentieth century, male adventure had become a vehicle for homosexual fantasy, and an inspiration for homoerotic art (drama and literature) (E. White 1993).[122] Casting doubt over the asexuality or heterosexuality of the adventure hero, readers such as Symonds, Lefroy and Genet contributed to the 'unmanning of manliness' (Vance 1985). Women, too, are sometimes able to enjoy the spectacle of male adventure, whether for the pleasure of looking at men, or for the satisfaction of seeing homoerotic adventure 'unmanning' its heroes and subverting the masculinity they embody. As film reviewer Lizzie Francke suggests, while 'all those hormones have to be seen to be coursing in the "correct" hetero direction' in adventure movies, 'the very male overdrive of the films always suggests otherwise' (Francke 1995: 26). Actively and critically reading (and enjoying) male adventures, from which she and other women are ostensibly excluded, Francke illustrates the agency of readers and the ambivalence of adventure texts, both of which facilitate critical readings.

Since they can be interpreted as critical narratives, adventure stories can function as critical narratives. Adventures which can be read, more specifically, as critical of British imperialism, also pre-date the now-famously 'anti-imperial' adventures of the late nineteenth century, such as *Heart of Darkness*. Adventure did not evolve from always supporting the *status quo* to become critical in the late nineteenth century. Earlier adventure stories can be interpreted as critical too. Zweig reads adventure stories, in general, 'as cultural myths in which the adventurer appears as a darkly anti-social character' (Zweig 1974: 16). He shows that adventure can be read as an anarchist story. Heroes such as Gilgamesh, Odysseus and Herakles ventured into the 'myth-countries', the 'world beyond men', and affirmed 'the possibility that mere men can survive the storms of the demonic world' (Zweig 1974: vii). Modern adventure, likewise, can be read as a literature of dissent. Marxist critic Michael Nerlich, for example, argues that Defoe was 'the next thing to a revolutionary' in early eighteenth-century Britain (Nerlich 1987: 260). Crusoe's journey leads him out of the contemporary British *status quo*, though a disorienting storm, into a seemingly uncomplicated, realistic, new-world space. The adventure becomes an anti-establishment, uncompromisingly bourgeois, non-conformist comment on, and utopian vision for, Britain. Defoe's adventurer, washed up with clothes, Bible, tools, language and other fragments of his 'civilisation', transplants selected elements of Britain to a setting where he can purify and nurture them, free from institutional constraints. His religion, in particular, is purified in the uncomplicated island setting where there is only a man with a Bible, and no Church of England. In addition to its domestic political criticism, *Robinson Crusoe* can be read as an attack on certain forms of imperialism, notably mercantile capitalism, slavery and plantation agriculture, which are relegated to the hero's 'wicked' youth. Ancestors of Defoe, whose adventure stories can also be read as radical narratives, include English and French anarchists and revolutionaries such as Henry Neville and

Gabriel de Foigny, whose work I discuss in this chapter. Reading Foigny, I illustrate the longevity of the radical tradition in adventure literature, which pre-dates the late nineteenth-century critical turn identified by readers such as Andrea White.

In some cases, readings of adventure as a critical narrative are dominant, while in others they are largely blocked, remaining latent or marginalised. The double life of adventure, the narrative which can be read as either critical or conservative (or both), is particularly evident in the works of Jules Verne, and in the genre in which he wrote – extraordinary or imaginary voyages (these terms are inter-changeable, Atkinson 1966). Modern adventure stories from *Robinson Crusoe* to the works of Jules Verne can be located within a tradition of critical extraordinary voyages, which can be traced from radical writers in revolutionary France, through eighteenth- and nineteenth-century European literature.[123] French extraordinary voyages, translated into many languages, were readily absorbed into international seventeenth-, eighteenth- and nineteenth-century adventure literature. In Britain, Verne was read as a conservative, pro-imperial writer by the dominant textual community led by critics and educators, who effectively blocked alternative readings. In particular, they blocked the possibility of a reading of Verne as anarchist and anti-imperialist, which contemporary critics show to be plausible. While this remains an alternative reading of Verne, it is the dominant reading of many comparable extraordinary voyages. The first of the modern 'radical' French extraordinary voyages, Gabriel de Foigny's *La Terre Australe connue* (1676), abridged in 1692 and translated into English as *A New Discovery of Terra Incognita Australis* (1693), has been labelled radical by the contemporary religious leaders and subsequent critics. While the radical reading of Verne was marginalised, the radical reading of Foigny was acknowledged. And while Verne was celebrated by critics, to the extent that popular writers ever are, Foigny was marginalised. In both cases, the dominant textual community sought to block readings of critical adventure, to deny readers access to a site of resistance.

A SPACE OF ANARCHY: THE OFFICIAL READING OF AN EXTRAORDINARY VOYAGE

The official reading of an extraordinary voyage, Gabriel de Foigny's *A New Discovery*, illustrates some of the mechanisms by which an adventure story may be read, hence the mechanisms by which its geography may come into cultural circulation, as a radical narrative. Key influences in the official reading of a geographical narrative such as Foigny's are contemporary opinion leaders, whose influence and authority may be formal or informal; contemporary and sub-sequent literary and cultural critics and reviewers; and the intentions of the author, whether the intentions are stated or whether they are inferred by others, often on the basis of biographical information about the author.

Gabriel de Foigny (1630–1692) wrote the 'first complete novel of extra-ordinary voyage' (Atkinson 1966: 163). The full title of his adventure story summarises the plot: *La Terre Australe connue: C'est à Dire, La Description de ce pays inconnu jusqu'ici, de ses moeurs & de ses coûtumes. Par M. Sadeur. Avec les avantures qui le conduisirent en ce Continent, & les particularitez du sejour qu'il y fit durant trente-cinq ans & plus, & de son retour* (1676). A second, abridged edition was published in French in 1692 and in English translation the following year, under the title, *A New Discovery of Terra Incognita Australis, or the Southern World, by James Sadeur, a French-Man who Being Cast there by a Shipwrack, lived 35 years in that Country, and gives a particular Description of the Manners, Customs, Religion, Laws, Studies, and Wars, of those Southern People; and of some Animals peculiar to that Place* (1693) (Plate 6.1). The English edition, which differs from the original in the details of its utopian philosophy, is broadly faithful to the original in setting and plot.[124] Sadeur, the hero, is a hermaphrodite, born at sea, who has many adventures at sea before he is finally shipwrecked and carried by flying monsters to the shores of Australia, a southern utopia, where he lives for thirty-five years. Set in Australia, and commonly regarded as 'the most famous of all the fictitious accounts of Australia' (Mackannes 1937: 156; Moors 1988), Foigny's adventure story can be located within the sub-genre of French extraordinary voyages set in Australia.

A New Discovery was branded a radical narrative, in the first instance, by the religious authorities where it was published. Foigny, in Geneva at the time, was called before 'the Venerable Company, created by Calvin as the guardian of the city's morals', which had been informed by two professors of theology that the work was 'full of extravagances, falsehoods and even dangerous, infamous and blasphemous things' (Berneri 1950: 186). Foigny was imprisoned and the authorities attempted to suppress publication and distribution of his book.

Foigny did not admit to the radical politics that others read into *A New Discovery* – he did not even admit to writing the book – but critics found evidence for the radicalism of *A New Discovery* in the radicalism of Foigny's life. Writers of extraordinary voyages were notoriously cagey about their intentions. One, unusually open about the purpose of his (and other) voyage narratives, explained that his story

> promises Amusement; it has all the ravishing Airs, and all the delightful Graces of a highly finished Romance; but at the same Time it is a severe and judicious Criticism, upon the almost innumerable Follies of the present Age.
>
> (Coyer 1750: iv)

Since writers were generally more guarded about their intentions, readers were left to guess, and often did so on the basis of biographical information about the person alleged to be the writer. No doubt the religious authorities, who imprisoned Foigny for writing an allegedly radical narrative, took into account his track record of transgressions. Other critics and historians read Foigny's story in the light of background information about the writer. They show how Foigny's biography as an anarchist and a master of geography is reflected in the

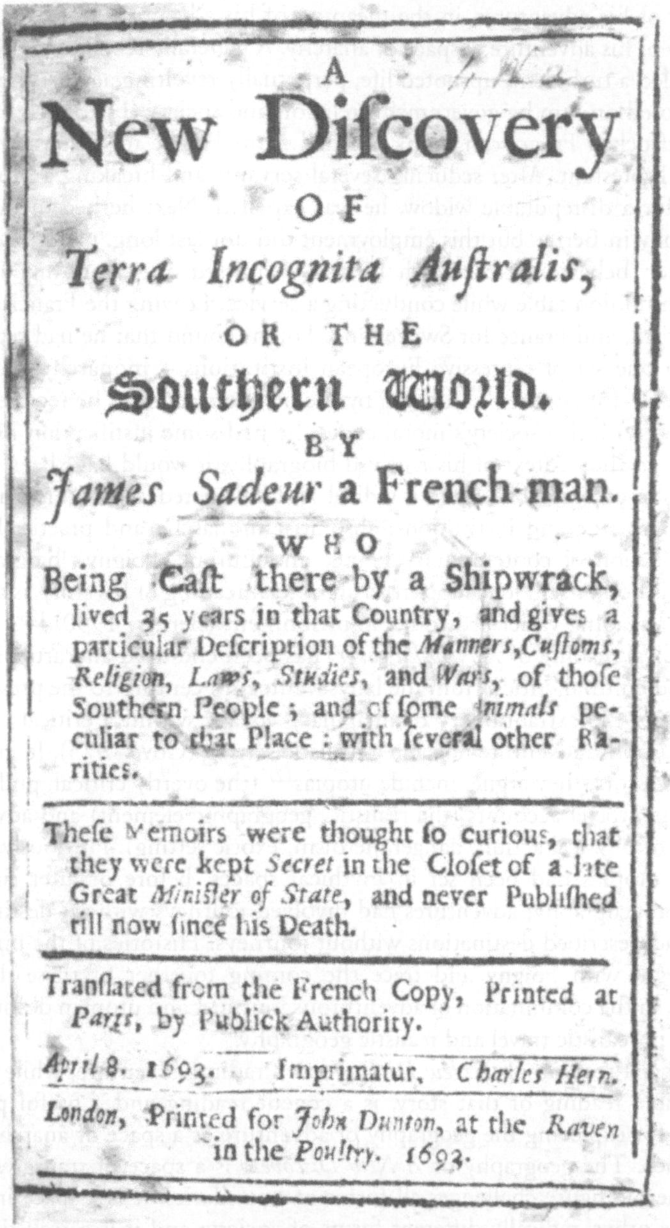

A

New Difcovery

OF

Terra Incognita Auftralis,

OR THE

𝕾𝖔𝖚𝖙𝖍𝖊𝖗𝖓 𝖂𝖔𝖗𝖑𝖉.

BY

James Sadeur a French-man.

WHO

Being Caft there by a Shipwrack,
lived 35 years in that Country, and gives a
particular Defcription of the *Manners*, *Cuftoms*,
Religion, *Laws*, *Studies*, and *Wars*, of thofe
Southern People ; and of fome *Animals* peculiar to that Place : with feveral other Rarities.

Thefe Memoirs were thought fo curious, that
they were kept *Secret* in the Clofet of a late
Great *Minifter of State*, and never Publifhed
till now fince his Death.

Tranflated from the French Copy, Printed at
Paris, by Publick Authority.

April 8. 1693. Imprimatur, *Charles Hern.*

London, Printed for *John Dunton*, at the *Raven*
in the *Poultry*. 1693.

Plate 6.1 Title page of *A New Discovery of Terra Incognita Australis* by Gabriel de
Foigny, abridged in 1692 and translated into English in 1693.

biography of his adventurer, in the meaning of his adventure story, and in the geography of his adventure, a space of anarchy. A generally rebellious, Bohemian figure, he led a turbulent, uprooted life, perpetually revolting against the restrictions imposed on him by government, religion and society (Friederich 1967). A young, defrocked Franciscan monk, he fled from France to Geneva where he became a Protestant. After seducing several servants and breaking a promise of marriage for a disreputable widow, he was expelled. Next he became a master of geography in Berne, but this employment did not last long, partly because of his drunken behaviour, for which he was dismissed after vomiting in front of the communion table while conducting a service. Leaving the Franciscans for Protestantism, and France for Switzerland, Foigny found that he had repeatedly exchanged one set of repressive European institutions – monarchy, patriarchy and Church – for another.[125] Judging by the hostile receptions he received when he (frequently) broke society's moral codes, he had some justification in feeling oppressed. In the context of his received biography, it would have been entirely characteristic of Foigny to write a radical and calculated assault on patriarchy, Christian doctrine and institutions, absolutist monarchy and practically every other institution of contemporary France and Europe. Foigny's biography, as much as his book itself, lent support to an official reading of his story as a radical narrative, a reading which led to his imprisonment (Berneri 1950).[126]

The radical reading of *A New Discovery* has been endorsed and articulated by literary and cultural critics, from the late seventeenth century to the present day. Critics locate the extraordinary or imaginary voyage within a critical tradition – the philosophic adventure novel in a realistic setting (Gove 1975). Its principal literary ancestors, they argue, include utopias[127] (the overtly critical, philosophical element), travel accounts (the realistic, geographic element) and adventures (elements of physical action, danger, heroism, exotic setting). Previously, adventures and utopias had been set in mythical spaces, before or after time and outside real geography; adventures had involved journeys without destinations; utopias had described destinations without journeys. Histories of the imaginary voyage begin with Foigny and trace the coming together of these disparate traditions, in the combination of adventurous journeys and utopian destinations, grounded in realistic travel and realistic geography.

The official reading of *A New Discovery* as a radical adventure, while not the only possible reading of that story, is a cogent reading and a useful point of departure for exploring the geography of adventure as a space of anarchy, a site of resistance. The geography of *A New Discovery* is a space of transgression in which a heroic figure challenges all forms of boundaries; it is a space in which to imagine other, radically different forms of society; and it is a realistic space, in which criticism is coded and naturalised in the form of realistic geographical imagery. *A New Discovery* is a comment on Europe, where Sadeur's life and journey begins and ends. With a French father and Germanic mother (named Willametta Ihn), he is generically European. European in style, too, in its Enlightenment reaction to the authority of many traditional institutions and rulers, *A New Discovery* (like other imaginary voyages) was a sweeping, Enlightenment vision of a brave new world, written from somewhere near the heart of the broadly defined *ancien régime*.

The geography of *A New Discovery* is a space of transgression in which a heroic figure challenges all forms of boundaries – spatial and social. Sadeur is an anti-social figure who wanders across society's boundaries, which cannot contain him. From the moment he is born – at sea, away from civilisation – Sadeur is outside society both literally (spatially) and metaphorically. 'Conceived in America' (Foigny 1693: 3), he is always a European outsider. The death of his parents, which occurs while he is still a baby and before he has reached dry land, seals his fate as a solitary wanderer. Born a hermaphrodite, 'he' soon finds that there is no place for him in European society, with its inflexibly gender-divided roles and its brutal sexual phobias. As a baby, orphaned at sea, Sadeur's 'matron' loses interest in caring for him when 'she found I was of two sexes', and soon she 'conceiv'd such an aversion for me, that it was a trouble to her to look upon me' (Foigny 1693: 10). Since everyone in Europe must become a man or a woman, he arbitrarily becomes a man, with a man's clothes, a man's name, and a man's pronouns. As a 'man' but also a hermaphrodite, Sadeur is perhaps only metaphorically hermaphrodite, which in Foigny's day meant (what we now refer to as) homosexual (Fausett 1993). Thus Sadeur transgresses European social constructions of both gender and sexuality. Out of place in Europe, he seems to belong more in a (contemporary) travellers' tale – in which hermaphrodites are relatively common – and that is where he must go (if not by choice).[128]

Sadeur continually transgresses spatial boundaries, and in his departures he symbolically rejects social orders, and the institutions and rules that go with them. His journey is marked by a series of violent departures. Sadeur's first storm and shipwreck ends in his deliverance into the world. His next sea journey is ended not by a storm, but by a Portuguese cannon ball, although Sadeur floats to safety in his cradle, and is taken on board the Portuguese ship. Several years later, when he is a prisoner on board a pirate ship,

> it began to blow terribly, and became so tempestuous that the Master pilots despaired of escaping; the Mast of our vessel broke, the rudder split, the ship leak'd on all sides, and we endured 24 hours the mercy of the Waves.
>
> (Foigny 1693: 17)

Once more Sadeur is saved. This time a Portuguese merchant ship finds him clinging to a floating plank of wood. Each of Sadeur's three early shipwrecks act as critical moments in his life, rites of passage in which he is violently removed from one condition, one chapter of his life story, and delivered into another. Early storms and shipwrecks prefigure the principal adventure, involving storms, shipwrecks and flying monsters, in which Sadeur finally, violently leaves all things European behind him. He is on board a Portuguese merchant ship, within sight of a port on the coast of Madagascar, when 'an East Wind so furiously tost the Sea, and drove us with that impetuosity that it broke our Cordage, and drove us above a thousand Leagues to the West' (Foigny 1693: 32). When his ship breaks in two, Sadeur catches a light plank and drifts alone, becoming disoriented.

> The Waves did so often plunge me under, and overturn me, that tho I held out as along as I could, yet at last I lost both knowledge and thinking, and truly I know not what became of me, nor by what means I was preserved from death; I only remember that coming to my self I opened my eyes and found a calm sea, I perceived an isle very near.
>
> (Foigny 1693: 33)

In most contemporaneous extraordinary voyages, the storm marks the traveller's passage from Europe to the other world, commonly a utopian or dystopian world. But for Sadeur, even after he has dragged himself onto the island and recovered, the ordeal is not over. Climbing a tree to get a view of the island, he encounters two 'prodigious flying Beasts' and a pack of other wild beasts, all of which chase him back into the ocean. After drifting further on his plank, Sadeur finally arrives on another island, which he calculates to be 35 degrees south latitude. Again, he is attacked by beasts, and is 'all bloody' when two of the 'great birds' appear, cutting him with their talons, before 'one of them seized me between her two feet and lifted me up very high in the air' (Foigny 1693: 41). Sadeur is not defeated. He fights the giant bird, tearing its eyes out with his teeth, and causing it to plunge into the water. The storm, the shipwreck and the encounter with flying beasts are images of crisis, of disorienting, violent, prolonged and painful departure. In the painful and disorienting journey, all the traveller can do is abandon himself to the storm, attempting to stay afloat and alive. He survives, but he is barely alive and he has lost the last traces of his old civilization – his clothes. Thus Sadeur leaves Europe completely behind him.

The violence that accompanies Sadeur's continual departures presents the departure from the European *status quo* as a violent – long, stormy and perilous – but necessary leap towards freedom. The disorienting and violent storm, a familiar literary image in early-modern Europe, presented an image of crisis. Some used the storm image as a warning of something that is best avoided. Edmund Burke, for example, was later to warn against the storm-like Revolution the French had experienced (Landow 1982). But Foigny presented the storm as inevitable, a difficult but necessary departure. In the storm-like crisis there was also opportunity to surface somewhere else, to invent something else. Literary critic George Landow generalises that the 'situation of crisis creates or generates an entirely new imaginative cosmos for those who experience it' (Landow 1982: 5). Since the new imaginative cosmos was unknown, Sadeur's adventure is primarily a journey *from* somewhere, not a journey *to* somewhere. It is a desperate leap into the unknown, in which the hero takes leave of his senses.[129] But there is life on the other side. When Sadeur is finally spotted by Australian guards, patrolling their coastline in a small boat, he is barely alive. His body is torn and bloody, and his clothes have all been shredded and ripped off. Disoriented and vulnerable, he represents humanity in a state of crisis, lost enough to emerge in some previously unknown space. All this works to his advantage. A naked hermaphrodite, arriving in a nation of naked hermaphrodites, and displaying great courage in the process, Sadeur is given medical treatment, fed, rowed to shore, and welcomed to Australia as a brother. 'I was in this Country, and amongst these *New faces, like a man fallen from the Clouds*' (Foigny 1693: 48). So, after his perilous adventure, the lost traveller finds that he has, indeed, been cast into an 'entirely new imaginative cosmos' – Australia. In Australia, Sadeur finds the appearance of utopia, but in his search for freedom he eventually (after thirty-five years) transgresses the boundaries not only of Europe, but of Australia too. Although Australia presents him with a vision of another world, beyond the *status quo* of contemporary Europe, he is ultimately too much of an anarchist to believe in the rational, ordered utopia that such an Enlightenment vision presents. At first, Australia seems to be a rational and free society, ordered but not ruled, where liberty comes from an 'adherence to strict Reason' (Foigny 1693: 78). There seem to be none of the formal authorities, such as the cumbersome monarchies and religious institutions, that rule and oppress Europe. On a general level, if not in its precise details, Foigny's Australia has much in common with the utopias of other seventeenth- and eighteenth-century European imaginary voyages. Sadeur seems to have found 'a perfect image of the state that man at first enjoyed in paradise' (Foigny 1693: 76). The adventure is suspended while Foigny recounts what he is told of the Australian utopia. But cracks begin to appear, as the European notices flaws in utopia, including vicious *urgs* (flying beasts) that Australians cannot control (despite their attempts to eradicate *urg* habitats), and warring *Fondins* (neighbouring tribes) that disturb the Australian utopia and lead the Australians to acts of rational but brutal violence. And even in Australia, the land of hermaphrodites, Sadeur is out of place, partly because he retains something of the old order; rationality has not entirely taken him over, and he remains capable of showing both compassion (towards *Fondins* – enemies whom the Australians kill) and passion (towards a young *Fondin*), for which he is sentenced to death. Sadeur manages to escape. Enforced rationality, Foigny suggests, would imprison the human spirit in the future, just as institutions of Church and monarchy had in the past and the present. One set of unfreedoms would be exchanged for another. The only solution was to keep running, transgressing boundaries and the unfreedoms they represent.

The geography of *A New Discovery* is a realistic space, in which criticism is coded and naturalised. Circumstantial realism, introduced to English readers by Defoe in the eighteenth century (Watt 1951), was already familiar to readers of French literature when Foigny wrote *A New Discovery*. The convention of *vraisemblance* – 'probability, or verisimilitude, in fictional works, over unbridled fancifulness' (Dunmore 1988: 23) – required that settings be plausible to contemporary readers, resembling places they knew first hand and geographies they knew from maps, exploration narratives and other travellers' tales. Sadeur's perilous journey is broadly realistic. Much of what happens in *A New Discovery* may appear fantastic to the modern reader, but by seventeenth-century standards it was realistic (Moors 1988).[130] There was nothing in Foigny's tale of pirates, storms and shipwrecks, flying beasts and hermaphrodites, that readers of ostensibly non-fiction, 'realistic' travel writing were not already accustomed to. Hermaphrodite nations were 'common rarities' in medieval travellers' tales and in imaginary voyages, and their existence had yet to be disproven. The 'Ruk' or giant flying bird survived from medieval travel tales and legends into the seventeenth century (Moors 1988). Foigny's tale, realistic as it was, could be read as a true story.

Writing a 'true' story, set in realistic space far away from Europe, Foigny spatially displaced his critique of Europe, disguising it from those who would censor or punish more direct criticism. As in all adventure stories, little is said of the adventurer's home, although everything that happens and everywhere the adventurer goes is to some extent a comment on that place. Although he is some-thing of an outsider there, Sadeur's home is in Europe, where his adventure begins and ends. The perilous journey is an act of leaving home, and Australia is defined in relation to that home. The rationality of Australia is contrasted to the irrationality of Europe. In the rational light of Australia, for example, Sadeur could see that there was no reasonable basis for the subjection of women and children to men in Europe. 'I found my self forc'd to believe that this great power which man had usurped over Woman, was rather the effect of an odious Tyranny, than a Legitimate Authority' (Foigny 1693: 72). Questioning the authority of 'the Father' (Foigny 1693: 71), and imagining a society without gender divisions of any kind – a society in which reproduction was a purely personal, private act, and in which domestic labour was eliminated by communal living and by a diet of fresh, uncooked fruit – Foigny took a broad swipe at European patriarchy. He also attacked European religious institutions, by showing how Australian morals were 'inspir'd by the light of Nature and Reason' (Foigny 1693: 73), and how Australian religion was a purely personal matter, how religious beliefs were left unspoken. An old man tells Sadeur that it is 'most certain that Men cannot speak of anything that's incomprehensible, without having divers Opinions of it' (Foigny 1693: 82).

Realistic, plausibly true stories such as *A New Discovery* sometimes get past the censors who would block the publication of radical political literature. Sadeur's eventual departure is abrupt and not very convincingly explained, but it is a necessary part of the story since it enables the 'editor' to claim the story is true (as such 'editors' conventionally did). Foigny's adventure was presented as genuine memoirs, which were 'thought so curious, that they were kept Secret in the Closet of a late great *Minister of State*, and never published till now since his Death' (Foigny 1693: title page). He concludes, 'These are the contents of Sadeur's memoirs written with his own hand' (Foigny 1693: 186). Thus he is able to disclaim authorship, hence to distance himself from the subversive act of writing political criticism. These were standard tactics of the imaginary voyage writer, as were attempts to preserve anonymity and to conceal details regarding places of publication and printers, by issuing books with false imprints, for example.[131] *Terre Australe connue* originally appeared under the false imprint of Vannes, France, with the false publisher, Jacques Verneuil. In some cases writers and publishers were successful in concealing their identities, although Foigny's rather half-hearted efforts[132] – perhaps he was parodying rather than following the convention – did not fool many people in Geneva, where he and his printer were soon identified (Berneri 1950). Nevertheless most imaginary voyages – even Foigny's – were received as true stories by at least some of their readers.

Realistic, lively stories appeal to popular audiences, hungry for geographical info-tainment, which puts them into cultural circulation and therefore establishes the possibility of some form of political impact. When political criticism is articulated in the form of imaginary voyages, it tends to reach broad audiences, people who are more likely to read lively, sensational stories like Foigny's than they are to read abstract political pamphlets. As one contemporary of Foigny observed, readers were 'ever delighted with strange accounts of other Countries, and purchase no books more than those upon such Subjects', which were 'all the reigning Taste' (Anon. 1757: 83). To write 'strange accounts of other Countries' was to ensure commercial success, on the one hand, and to retain the possibility of political effectiveness, on the other. Both prospects appealed to Foigny, who wrote *A New Discovery* partly to make money and partly to voice his political views. Since the publication of Foigny's story (like other extraordinary voyages) was shrouded in secrecy, it is difficult to trace its publishing history – numbers of copies printed and sold, for example – although the reprints and translation qualify it as something of a success, as do reports that it caused an immediate sensation when it was published (Atkinson 1966). It therefore had the potential to reach relatively wide audiences, although the question of what they made of such a sensational, ambiguous, diffuse narrative is not easy to answer. But, judging by their actions, the Venerable Company of Geneva must have expected general readers to understand *A New Discovery* as a radical, anarchistic adventure story, a threat to society.

The geography of *A New Discovery* is a space in which to imagine other, radically different forms of society. Foigny, like other writers of utopian and dystopian adventurers who have sent heroes to *terra incognita* – fantastic, imaginary spaces and space that is 'real' but unknown (to Europeans) – used the setting of his adventure as a conceptual space in which to reflect upon, and perhaps reinvent, Europe. Foigny's setting – Australia – had, since the beginnings of the modern period, constituted the largest blank on European maps (Plate 1.3). From the seventeenth to the twentieth century, no *terra incognita* was hazier or more pronounced on European maps than Australia. Early in the century Bishop Hall, a British satirist, commented that Australia was conventionally regarded as real but unknown. In the voice of a would-be adventurer, he reflected that

It hath ever offended mee to looke upon the Geographicall mapps, and finde this: Terra Australis, nondum Cognita. *The unknowne Southerne Continent. What good spirit but would greeve at this? If they know it for a Continent, and for a Southerne Continent, why then doe they call it unknowne?*

(Hall 1609: A3–A4)[133]

This 'unknown' continent was to serve as the setting of many extraordinary voyages. *A New Discovery* was 'the first English-printed work in which this continent is distinctly named Australia' and 'probably the first in any language in which Australia is made the subject of an imaginary voyage and the scene of a fanciful Utopia' (Blair 1882: 204).[134] As such, it is an important ancestor of imaginary voyages such as *The Isle of Pines* (1668) by Henry Neville (set on an island off the coast of Australia); *The History of the Sevarites* (1675) by Denis Vairasse D'Allais; Jonathan Swift's *Gulliver's Travels* (1726); Peter Longueville's *The Hermit* (1727); Abbé Coyer's *A Discovery of the Island Frivolia* (1750); anonymous works such as *The Voyages, Travels and Wonderful Discoveries of Captain John Holmesby* (1757); and nineteenth-century descendants including *Swiss Family Robinson* (Wyss 1814)[135] and stories set in the Australian interior by such writers as Lady Mary Fox (1837), Ernest Favenc and Jules Verne. Australia, as a setting of adventure, provided conceptual space in which to think critically about European or colonial society, in Foigny's case the former.[136] Imagining something other than the *status quo*, *A New Discovery* was an attack upon the European *status quo*.

While Foigny coded and naturalised revolutionary politics in realistic images of a perilous journey, he coded reflections on the possibility of a different, better, post-revolutionary state in realistic images of Australia. The realistic geographical imagery of Sadeur's journey continues when he arrives in *Terra Incognita Australis*.[137] Sadeur sets out the broad outlines of Australia, which he says is approximately 3,000 leagues in length and 500 in width.

I have here . . . set down the best account of the *Australian Territories* that I could get either by the relations of others, or cou'd describe according to the Meridian of *Ptolemy*.

It begins in the three hundred and fortieth Meridian, towards the fifty second degree of Southern Elevation, it advances on the side of the Line, in forty Meridians, until it comes to the fortieth degree. The whole Land is called *Hust*: The Land continues in this elevation, about 15 degrees, and they call it *Hube*; from the fifteenth Meridian the Sea gains, and sinks by little and little into twenty five Meridians, until it comes to the fifty first degree. And all on the western side is called *Hump*: The Sea makes a very considerable Gulph there, which they call *Ilab*: The Earth afterwards falls back towards the Line, and in four Meridians advances unto the two and fortieth degree and a half; and this Eastern side is called *Hue*: The Earth continues in this elevation about thirty six Meridians, which they call

Huod; after this long extent of Earth, the Sea regains, and advances unto the forty-ninth degree, in three Meridians, and having made a kind of a semicircle in five Meridians, the Earth returns and goes on unto the thirtieth degree, in six Meridians, and this Western side is called *Huge*. The bottom of the Gulph *Pug*, and the other side *Pur*; the Land continues about 34 Meridians, almost in the same elevation, and that is call'd the Land of *Sub*, after which the Sea rises, and seems to become higher than ordinary, wholly overflowing the Earth, and falls again by little and little towards the Pole, the Earth by degrees giving way unto the sixtyieth [*sic*] Meridian, on this side are the Countries of *Hulg, Pulg*, and *Mulg*; towards the fifty fourth degree of elevation, appears the mouth of the River *Sulm*, which makes a very considerable Gulph. . . .

Thus the Australian Territories contain twenty seven different Countries, which are all very considerable, and are altogether about three thousand Leagues in length, and four or five hundred in breadth.

(Foigny 1693: 48–51)

This description echoes both the style and the content of contemporary geographies. The style is plain; Sadeur tirelessly lists geographic details until the reader is wearied into believing them. The content of Sadeur's description is in broad agreement with contemporary geographies of Australia, which were still very sketchy. As one of Foigny's contemporaries explained in 1675, in the introduction to his own imaginary voyage,

Among all remote Countries, there is none so little known, and so vast, as the third Continent, commonly called Terra Australis. It is true, Geographers have given some small and unperfect descriptions of it, but it is with little knowledge and certainty; and most of the draughts may be suspected, and look'd upon, as imaginary and fictitious. Sure it is, that there is such a Continent; many have seen it, and even landed there, but few durst venture far in it, if any there were; and I do not think that any body hath made any true description of it, either for want of knowledge, or other necessary means and opportunities.

(Vairasse 1675: A5)

Throughout the seventeenth century, Australia was little more than the generalised and incomplete outline of a huge continent, both on most maps and also in the geographical imaginations of most Europeans. The Australian coast was not accurately and systematically charted until after Foigny's story was published.[138] During the seventeenth century a number of Dutch navigators encountered Australian coastal waters, and some of their experiences were published and incorporated into contemporary maps, consolidating the popular image of Australia as a large but mostly unknown land.[139] That general image dated back to at least the second century AD, when Ptolemy suggested the existence of a vast southern continent (Gerrard 1988), and it was only reinforced

127

by the tales and legends of medieval travellers (from Sinbad the Sailor to Marco Polo), then by uncertain French and 'secret' Portuguese 'discoveries' of (what became) coastal Australia (Cowley 1988). The first relatively formal narratives of 'Australian' exploration were Spanish, published in the first two decades of the seventeenth century, although their author mistook the New Hebrides for Australia, and only described limited areas, nothing like a whole continent.[140] So when the Dutch began to chart the coast of Australia later in the same century, they were charting a coast that was virtually unknown to most of their fellow Europeans.

Foigny used unknown Australia – its mostly unknown coast and its almost entirely unknown interior[141] – as *carte blanche* in which anything was possible. He wrote as if Sadeur were the first European to encounter Australia and live to tell the tale. Earlier travellers, he claimed, 'have either been lost in their voyage, or have been killed by the inhabitants of the country after they had entered it' (Foigny 1693: A2). He dismissed the reports of earlier travellers, from Magellan and Marco Polo to Queiros. 'Tis therefore to our Sadeur, whose relation here follows, that we are wholly obliged for the Discovery of this before unknown country' (Foigny 1693: A2). This gave Foigny complete freedom to invent Australia out of the open space framed by an incomplete and hazy outline. He extends realistic travel and realistic geography into the Australian interior, seamlessly blending received geographic 'facts' with his own speculations, and maintaining his plain, realistic, descriptive style throughout.

Like the *terra incognita* on European maps, Foigny's Australia is never very real; it is conceptual space in which Europeans think about Europe. It is completely uniform in topography and soil fertility, as in everything else. 'This great country is plain, without forests, marshes, or desarts, and equally inhabited throughout' (Foigny 1693: 59). There are no seasons, no clouds and no rainfall, although rivers ensure a sufficient water supply. In Foigny's Australia 'there is neither flyes, nor caterpillars, nor any other insect. There's neither spider, nor serpent, nor any venomous beast to be seen' (Foigny 1693: 62). Human settlement is uniform and symmetrically ordered, as are languages, customs and buildings. Sadeur learns some of this from what he sees, but more from what he hears, and his descriptions reflect that detachment. He describes Australia with grand perspective, as if from a great height. Mainly, Foigny's Australia is a conceptual space in which to imagine a possible world. A very abstract geography, it has more in common with the isotropic plains of twentieth-century spatial-scientific geographers like Walter Christaller (who devised a theory of settlement patterns) and Alfred Weber (who modelled industrial location) than it does with any recognisably or plausibly real geography.[142] The prospect of *carte blanche* in the Australian interior, away from the cluttered realities and crowded maps of contemporary Europe, inspired Foigny to invent his sweepingly rational, logically ordered utopia.

Although *A New Discovery* is 'officially' read as a radical adventure story, it could be read otherwise. It could, for instance, be 'just' a story, a sensational story at that. At least some readers of Foigny's tale, as of other imaginary voyages, thought they were reading a true story, and there is no need to accept the 'official' view that this reading was naive; it was just another reading. *A New Discovery* could, alternatively, be read as a story about Australia rather than Europe. Extraordinary voyages tend to lead double lives, as comments on Europe, on the one hand, and as comments on their settings, on the other. Most could be read in both ways. For example, although *Voyage de Robertson aux Terres Australes* (Anon. 1767) was intended as an attack on the French government, it is also 'supposed to have inspired William Penn to found an ideal settlement in North America' (Moors 1988: 13). There are surely additional, possible readings of *A New Discovery*, not acknowledged in the 'official' reading. But, probably because it is relatively obscure, the meaning of Foigny's adventure story has never been seriously disputed. In contrast, many readers of another writer of French imaginary voyages, Jules Verne, have disputed 'official' or dominant meanings. Two textual communities have constructed 'two Jules Vernes', one of them an anarchist like Foigny, the other a conservative.

DOMINANT AND ALTERNATIVE READINGS OF JULES VERNE

The dominant reading of Jules Verne, particularly in Britain, was and is one of an imperialist and a conservative. The interpretation of Verne, like that of Foigny, has been guided by opinion leaders, biographers and literary critics.

Jules Verne (1828–1905) made a strong impression on British readers, who helped make him the best-selling author of all time (Butcher 1990). As a writer of adventure fiction, his cultural significance in Victorian and Edwardian Britain was second to none. While *Robinson Crusoe* remained the most important adventure book, Verne became the most important adventure writer. His British following was partly a response to his penchant for British heroes and British Empire settings. Verne's adventure stories, dominated by a series he called the extraordinary voyages, were written in French and originally published in France, but were quickly – often too quickly, hence poorly (Butcher 1990) – translated into English. They were then readily absorbed into English-language adventure literature (and other media such as cinema). Contextualising Verne's stories, from the perspective of Britain, I look not to first editions and French histories and geographies, but to English editions and British readers.

The extraordinary voyages began with *Cinq Semaines en ballon* (1863) and *Voyage au centre de la terre* (1864), closely followed by *De la Terre à la Lune* (1865), *Voyages et aventures du Capitaine Hatteras* (1866a, 1866b) and *Les Enfants du Capitaine Grant* (1867–1868), and another fifty-nine titles, approximately one per year until 1910 (the last few posthumous). *Les Enfants du Capitaine Grant*, translated into English and published in three parts under the general title *A Voyage Round the World* (1877), is illustrative of Verne's extraordinary voyages, although no single title can be called representative because the voyages are complex and varied (Butcher 1990). It is, to summarise, the story of a quest in search of Captain Grant, who has been wrecked with two seamen while on a mission to establish a New Scotland in the Pacific. A message in a bottle alerts Scottish aristocrat Lord Glenarvan, who mounts an expedition in search of Grant, accompanied by a party including Lady Glenarvan, a French geographer named Paganel and the children of Captain Grant. Aboard the Glenarvans' private yacht, *The Duncan*, the party sail to Patagonia, Australia and New Zealand, on their continuing search for Grant. The partly British cast and British Empire settings of *A Voyage Round the World*, and its popularity in Britain and colonial Australia (Faivre 1969), illustrate Verne's contribution to 'British' adventure fiction. And its similarities to *A New Discovery*, as a French extraordinary voyage to Australia with many of the same images and adventure motifs, makes for an interesting comparison. How, I must ask, could the two very similar adventure stories be read so differently? This question may be answered historically, with reference to readers and the web of power relations in which readings take place. Readers with particular influence in the interpretation of Verne, as of Foigny, include biographers, opinion leaders and critics. For around a hundred years after the publication of his first books, these readers clung to a single image of Verne and his fiction. Despite the heterogeneity in Verne's *oeuvre*, which some critics have emphasised recently (Butcher 1990), readers traditionally generalised and simplified Verne.

Biographies and obituaries of Verne traditionally confirm the dominant, conservative reading of the writer and his work. *The Boy's Own Paper*, which placed him 'among the greatest of France's literary sons', presented Verne as a man of adventure, in keeping with the best *BOP* traditions. The popularity of Verne, like that of Ballantyne and other adventure writers, was fuelled by images – presented in biographies, obituaries and public appearances – of the author as a 'real-life' hero. Verne's obituary, printed in the *BOP* in 1908, took the form of an abbreviated adventure story. The reader learns that Verne was born on the banks of the Loire, the river that provided his first glimpses of the outside world.

> From earliest childhood, therefore, Verne was in contact with the life of the sea, and M. Charles Lemire, the author's biographer, tells us that it was one of the delights of young Verne's life to watch the ceaseless tide of shipping in the great estuary where, in after years, he was wont to moor his own beautiful yacht
>
> (Aitchison 1909: 558)

Thus a French biography of the author was filtered through to British readers, including boys, guiding their reading. The *BOP* goes on to tell the tale of how Jules, aged 11, stowed away on a boat bound for the Indies, only to be retrieved by his father. The obituary tells how, like other adventurers, Verne possessed a heated geographical imagination, and spent many hours gazing at maps and globes, and sketching maps of islands and other imaginary places, which accommodated his fantasies. According to the *BOP*, Verne's geographical fantasies were also inspired by the Anglo-American adventure works of Walter Scott and James

Fenimore Cooper, and by *Robinson Crusoe*, which 'afforded him great delight' (Aitchison 1909: 558). Biographers also emphasise Verne's stated intention, first announced in the fourth volume of extraordinary voyages, to summarise the whole of human knowledge: geographical, geological, physical and astronomical. Readers of Verne have traditionally accepted this version of his biography and allowed it to guide their interpretation of his works. For example, I.O. Evans presented Verne as an adventurer of the *BOP* variety in the series of extraordinary voyage editions he abridged, edited and introduced in the 1960s, and in his book on Verne, *Jules Verne and his Work* (1965). Readers of Evans' edited editions learn, before reading the stories, that Verne was a would-be stowaway who loved adventure stories. More serious critics use similar evidence to justify their reading of Verne. Attempting to refute a 1970s reinterpretation of Verne's *Twenty Thousand Leagues Under the Sea* as an anarchist novel, the London *Times Literary Supplement* (1972: 1391) quoted Verne as saying that 'politics has no place in the book'. Although, like Foigny, Verne worked within a genre associated traditionally with a coded radicalism, in which authors denied political intent, his stated intentions are accepted, and allowed to guide interpretations of his work. Thus, the dominant reading of Verne is supported by the biographical construction of the author as a man who loved excitement and faraway places, and was uninterested in politics.

British readings of Verne have also been guided by opinion leaders including reviewers, newspaper and magazine editors, educators and juvenile literature critics. Authorities on children's literature such as Edward Salmon praised Verne for the educational content of his stories. Verne was received in Britain as a popular geography teacher. The *BOP* explained to readers that his writing was preceded with periods of scientific and geographical study, rather like the preparation a boy reader might be advised to undertake before writing an essay.

> After a long course of preparation, involving a deep study of science and geography and the perusal of the voluminous notes he had taken as a boy from books, papers, and magazines, he produced a *History of Exploration*. The knowledge gained in the preparation of this work induced him to proceed, and in 1863 appeared the first of the extraordinary series of the *Voyages Extraordinaires*.

'His books', the *BOP* continues, 'were usually planned by means of a globe and maps and charts innumerable' (Aitchison 1909: 557). To the *BOP*, as well as to influential critics such as Edward Salmon, whose views on children's literature were published in journals such as *The Nineteenth Century* and in books such as *Juvenile Literature As It Is* (1888) and *The Literature of the Empire* (1924), Verne was a respectable writer who could be accommodated within a conservative, imperialist tradition of British adventure literature.

Compounding the views of biographers and contemporary opinion leaders, British literary critics have traditionally labelled Verne as a typist rather than a writer, a teller of tales rather than an author of literary works, and as a result they have associated him with storytelling tradition rather than literary creativity, with conservatism rather than subversive invention. They have done this, in part, through silence. By (all but) ignoring Verne, they imply that he is unworthy of serious critical attention, and refuse to accept him into the fraternity called Literature. Orwell (1940) wrote off Verne as 'unliterary'. Verne's extraordinary voyages have 'suffered from a reputation for naivety and poverty, and from at best an admittance into a sub-literary genre', and this reputation has discouraged and coloured the few academic readings of Verne in Britain (Butcher 1990: 163). As late as 1972, British critics suggested that if Verne must be read as literature, it should be as derivative, Romantic literature (*Times Literary Supplement* 1972: 1391). In the shadows of history and literature, Verne keeps good company; others such as Bessie Marchant still await the attention of biographers and 'serious' critics. As recently as 1990 a publisher could market a new book on Verne with the claim that

> In Britain . . . no full-length scholarly study of Verne has appeared to date.
> It is this remarkable gap, where the best-selling author of all time – and the only Frenchman to have achieved truly universal renown – is either completely unknown or travestied, that Dr Butcher brilliantly fills here.
>
> (Butcher 1990: dust jacket)

Macmillan's marketing people exaggerate a little; critical re-readings of Verne began to appear on the British market in the 1970s, particularly after *The Political and Social Ideas of Jules Verne*, by Jean Chesneaux, was published in English translation in 1972. But, as bibliographers Gallagher, Mistichelli and Van Eerde (1980: xvi) observed, 'English language criticism on Verne' had still not passed 'beyond a sterile and superficial level' in 1980. By their silences, and occasionally by their attentions to Verne, a century of British critics cemented the image of Verne's work as juvenile literature of the *BOP* variety: simple, boyish, conservative and imperial.

Verne's stories have been received as simple, straightforward stories in Britain, where he was immediately characterised as a *Boy's-Own-Paper* kind of writer. Indeed, he was popular among *BOP* readers. Charles Welsh's 1884 survey identified him as the fourth most popular author among schoolboys, considerably less popular among girls (Table 3.1; Salmon 1888). Evans, welcoming a revival of interest in Verne, suggested that the straightforward, heroic simplicity of the extraordinary voyages explained their appeal. He argued that

> the reading public, surfeited with mass-produced thrillers and westerns and 'space opera', with 'debunking' biographies, with 'sick' humour, the 'anti-hero', and the 'kitchen sink', are returning with relief to what is traditionally one of the great themes of literature, the story of strong, adventurous, and self-disciplined heroes and heroines, displaying dogged determination and stalwart courage in facing hardship and peril for the sake of a worthy ideal.
>
> (Evans 1965: 9)

Thus Verne is read as a map-maker, whose stories chart geographies and identities in bold colours. In the words of one modern critic, his 'best stories work very well at that boy-scout level of a group of male friends in an exciting mapping venture' (Suvin 1974: 59). Thus British (and other English-language) readers have received the extraordinary voyages as lively boys' stories, 'merely good fun' (Costello 1978: 16).

Verne's reception in Victorian Britain as a writer of juvenile literature was fuelled by the educational component in his stories, and by their amusing and entertaining style. While Verne, like most other writers of the later nineteenth and early twentieth century, toned down the didactic element in his work, particularly with regard to religious matters, he emphasised the factual dimension. Verne as teacher of facts was received as a writer for children. More specifically, he was received as a children's geography teacher; Verne was, indeed, an encyclopedic geographer (Martin 1985), a writer of geographical texts who probably reached wider audiences than any geographer before or since (with the possible exception of Defoe). In Australia, *A Voyage Round the World* has been described as 'a rich textbook' (Stevenson 1958). The geographical content of Verne's adventure stories was not lost on Fellow of the Royal Geographical Society I.O. Evans. Introducing *Captain Grant's Children* – his version of Verne's tale, with a more literally translated title – he draws readers' attention to the story's 'background of factual knowledge' and to the role of the geographer Paganel, a character who reels off factual information, educates his travelling companions and also his readers (Evans 1964a: 8). Indeed, *A Voyage Round the World* introduced nineteenth-century readers to Patagonia, Australia and New Zealand, as well as many other islands, in the company of a loquacious French geographer, whose geographical scholarship has won him 'a most distinguished place among the *literati* of France' (Verne 1877: 1.63). Through the voice of the narrator, and through the voice of Paganel, Verne assumed the tone of a geography teacher.

Verne's stories, characterised as juvenile and simple in their world view, have also been read as conservatively confident in the *zeitgeist* of Victorian Britain: confident in progress, enthusiastic about science and committed to imperialism. British readers have seen Verne as a scientist, a lover of science and technology, an inventor of science fiction (Evans 1965; Costello 1978). In English-language criticism he has been characterised as unreservedly pro-science, faithful in technology and uncritically confident in western material improvement (Gallagher, Mistichelli and Van Eerde 1980: xiv). In Verne's confident scientific vision, it is said, man always conquers nature, extending his imperial authority. Since he was first published in Britain, Verne has generally been regarded as a defender of empire, racist and masculinist – even misogynist. The ubiquitous maps and globes, which appear not only in Verne stories but also in film adaptations, can be interpreted as imperial visions. Culture critic Ella Shohat reads the cartographic images, which appeared in a film version of *Around the World in Eighty Days*, as imperial maps, 'linking the development of science to imperialist control over much of the globe, which thereby becomes unproblematically available for the scientific conquest' (Shohat 1991: 48). Typically, in *A Voyage Round the World* the adventurers imaginatively map the Southern Hemisphere, following the thirty-seventh parallel, as it appears on their maps, and bringing the world within its abstract cartographic, imperial vision. The world they encounter falls within their pre-defined, Eurocentric and imperial world view. The people they meet fall within this vision too. Aboriginal peoples conform either to the Romantic's noble savage (Thalcave, in Patagonia) or the Darwinist's primitive man stereotype (Australian aborigines), while white colonists (such as farmers in South Australia) signify the prospects for future progress. Verne's imperialism, like that of other Victorian adventure writers such as Haggard (Katz 1987), appears crudely racist. Australian aborigines, for example, are described as 'the lowest type of humanity' in *A Voyage Round the World*. Evidently, Verne's imaginative geography is located firmly within the British Victorian mainstream, as it is usually understood.

In recent years the dominant interpretation of Verne, which I have outlined, has been countered with alternative readings, which draw attention to the power of readers to construct meanings and chart geographies, in the open-ended cultural space created by Verne. Alternative readings of Verne, like the traditional readings, are the work of literary critics and revisionist biographers, in the first instance. They present a reading that remained latent or marginal in Britain until the 1970s, although it is by no means a new way of reading extraordinary voyage novels. Extraordinary voyages have been read as anarchist novels, their geographies as sites of resistance, since Foigny wrote *A New Discovery*, perhaps for longer. Had the alternative reading of Verne gained currency among his many British Victorian and Edwardian readers, Verne's reputation and his cultural impact would have been very different.

The re-reading of Verne began with revisionist biography, in the context of a critical tradition in which authors' intentions, motivations and characters were considered important insights into the meaning of their work. Jean Chesneaux went to great lengths to rewrite the biography of Verne, in order to establish his case for a re-reading of Verne. Behind the 'bourgeois facade', he argued, there lay a 'secret' Verne, an anarchist guided by the spirits of the 1848 revolution: utopian socialism and libertarian individualism. Chesneaux, like other critics (Compère 1974), searched for biographical evidence that Verne was an anarchist, and found some (although others have contested this 'evidence', for example Day 1967). They looked for slips in his behaviour and evidence of his secret self, which they also found. Verne's handwriting, for example, revealed 'traits of character far removed from ordinary bourgeois mentality: a lack of proportion, a certain secretiveness, a liking for solitude and great strength of will' (Chesneaux 1972: 19–20). There are problems with this argument, not least the assumption that a bourgeois cannot be a revolutionary, an assumption confounded by Defoe. But, more seriously, the assumption that the author's intentions must define the meaning of his or her text is untenable. Verne's intentions were never clear, but even if they had been they could not have determined the meaning of his stories, particularly outside France, where his immediate political interests lay.

Literary critics have challenged the dominant reading of Verne's stories, and the dominant literary mapping of Verne's geography, proposing the alternative reading of his work as serious literature and anarchist, anti-imperialist criticism. One critic recently insisted that Verne, understood so well (or so it seemed) by boys and marginalised so confidently by literary critics, remains 'unexplored territory' (Bradbury 1990: xv). Thus Bradbury prefaced a foray into the cultural space opened by Verne's writing, a preliminary remapping of Verne's stories and geographies. Bradbury prefaces a literary reading in which Verne's work is reinterpreted as Literature. The author remarks, for example, how 'Verne's prophetic literal-mindedness topples over into poetic ecstasy' (Butcher 1990: 167). For the most part, British critics have yet to remap Verne (as their French counterparts have done), whether as a literary figure or as a social critic (Gallagher, Mistichelli and Van Eerde 1980). They have, however, destabilised the traditional reading of Verne, denaturalised the geographies of Verne, and opened cultural space in which to re-read and remap. Briefly suggesting an alternative reading of Verne (indebted to critics such as Chesneaux), again with illustrations from *A Voyage Round the World*, I illustrate possible readings of Verne's geography that were, in effect, circumscribed by the powerful textual community comprised of opinion leaders, biographers and critics in Victorian Britain.

The geography of Verne's adventure can be read as a space of anarchy. Like other writers such as Defoe and Ballantyne, Verne littered his tales with geographical facts, not (just) for the sake of his readers' geographical educations, but as the medium for broader forms of instruction and suggestion. Whereas Ballantyne's geographical facts can be interpreted as the medium for moral instruction relating to manliness (Chapter 3), Defoe's for spiritual (and other) reflection (Chapters 2 and 7), Verne's geography can be read as the medium for political suggestion. Verne's realistic geography has much in common with the realistic geography of other extraordinary voyages that critics have labelled anarchist. The spaces of *A Voyage Round the World*, for example, parallel those of Foigny's *A New Discovery*. The settings of both novels are made to appear exotic, and they provide the European hero(es) with many of the same adventures, including encounters with forms of primitivism, among natives and wild animals. In both stories, for example, the Europeans meet 'primitive' people (cannibals in Verne, hermaphrodites in Foigny) (Fausett 1993), and in both stories they are attacked by giant birds of prey, which attempt to carry them away. Verne's giant bird is a giant Condor, illustrated in Plate 6.2, which snatches Robert Grant in its claws and soars to a height of several hundred feet before being shot down. In both Verne's and Foigny's stories, readers are lured by exoticism and excitement, and reassured by the factual appearance of a 'true' story, and, once they have entered the space of the text, they are open to its political suggestions. Verne's geography, like Foigny's, is linear, marked by continual violent departures. The adventurers continually escape, whether from the claws of a bird, the threat of a storm or the cooking pot of cannibals. Their continual departures, excused in Verne's case by a half-hearted quest in search of

Plate 6.2 The South American birds that attack and fly away with humans in *A Voyage Round the World* (1877) are reminiscent of the flying beasts in *A New Discovery of Terra Incognita Australis*, illustrating continuity between seventeenth- and nineteenth-century French extraordinary voyages.

a missing seaman, can be interpreted as continual, violent rejections of social order, transgressions of spatial and social boundaries. The true hero of *A Voyage Round the World* is the geographer, who like Sadeur is a wanderer, a social outsider who is most at home in the liminal geography of the road. Paganel is a free-wheeling lover of freedom (Faivre 1970), whose many hours spent studying maps give him access to faraway lands and seas, and whose geographical fantasies lead him continually towards *terra incognita*, the true setting of adventure. While Paganel is a gentle rebel, his story less violent than that of Sadeur, for example, he is nevertheless an anarchist visionary. Some of Verne's other adventurers are more explicit in this respect, most notably Captain Nemo, who sails under the black flag (of anarchy, perhaps) and insists upon living outside society, particularly the society of the British, who have killed his family in India (see Chesneaux 1972; Taussat 1974). Nemo's specific references to British India, like Paganel's to British Australia and New Zealand, where *A Voyage Round the World* is partly set, allow for a reading of Verne's stories not as a purely general anarchist vision, but as a specific attack upon British imperialism.

The geography of Verne's adventure can be read as a site of resistance, including resistance to British imperialism. In many Verne novels resistance movements are explicitly and sympathetically described. Readers of Verne's extraordinary voyages learn about resistance to the British Empire by French Canadians, Irish rebels and aboriginal peoples in colonised countries such as British Canada and Australia (see Chesneaux 1972). But, as a form of resistance to empire, the description of resistance movements is relatively cautious and unsubtle.[143] Cautious, because the European writer, narrator and hero(es) externalise the resistance, at most offering sympathy towards the colonised people who fight back against the British. Unsubtle, because the site of resistance is simply the visible setting, not the wider cultural space created by the story. But other forms of resistance to British imperialism, less cautious and more subtle, appear throughout Verne's extraordinary voyages. In *A Voyage Round the World*, Paganel's is the clearest voice of resistance, and it is the voice of a European, perhaps Verne himself. Paganel speaks out against the British Empire, particularly when he is in Australia, where he attacks British imperial ideology and British colonial violence.

Attacking British imperial ideology, Paganel gently teases an Australian Aborigine for his Anglo-centric geographical education. The boy, named Toliné (Plate 6.3), has won first prize (a Bible) for geography at the Melbourne Normal School, and when he realises that he is talking to a distinguished geographer he insists that Paganel 'examine' him. Asked to describe the principal divisions of Oceania, Toliné answers

> 'Australia, belonging to the English; New Zealand, belonging to the English; Tasmania, belonging to the English. The islands of Chatham, Auckland, Macquarie, Kermadec, Makin, Maraki, are also belonging to the English.'

'Very Good, and New Caledonia, the Sandwich Island, the Mendana, the Pomotou?'

'They are islands under the Protectorate of Great Britain.'

(Verne 1877: 2.156)

Paganel protests that they are French. But Toliné imagines the whole world as Britain's Empire. Even France is a province of England, with Calais the capital. Paganel laughs at Toliné's knowledge, which he calls 'Anglo-coloured fanciful geography' (Verne 1877: 2.157).

So that's the way they teach geography in Melbourne! . . . Europe, Asia, Africa, America, Oceania, the whole world belongs to the English. My conscience! With such an ingenious education it is no wonder the natives submit.

(Verne 1877: 2.159)

Paganel tries to undermine the imperial ideology encoded in Anglo-centric geographical imagination, subverting it through the use of humour and caricature, although he acknowledges that it is a stubborn ideology. Toliné takes no notice of Paganel's reaction, and does not accept the geographer's gift of an alternative geography book. Paganel learns that unmapping and remapping imaginative geography is neither an instant nor an easy process.

Paganel also attacks British colonial violence. In Australia, he speaks out against British brutality with respect to Aborigines, who have been herded like animals onto reserves and degraded through rough handling. He articulates the horror of Aboriginal reserves where natives

have been brutally driven by the colonists. In the inaccessible bush of these distant plains there are marked out fixed places where the aboriginal race may gradually become extinct. Any white man – colonist, emigrant, squatter, bushman – may cross the limits of these reserves – the black man alone must never leave them.

(Verne 1877: 2.188)

Paganel describes the 'sad spectacle' of a native encampment in the Murray Territory, where Aborigines, 'degraded by misery', were 'repulsive' to look at. 'Nothing was more horrible than their hideous faces, with enormous mouths, broad noses' and 'projecting lower jaw', and 'Never did human beings so closely approach the animal type' (Verne 1877: 192). Verne's grotesque Aborigines, rather than conventionally racist stereotypes (the traditional reading of Verne), can be read as victims of British conquest, their degradation more a reflection on the British than on themselves. Verne's anti-imperialism is most angrily explicit when Paganel accuses the British of genocide. This is worth quoting at length.

[Paganel] descanted on the difficult subject of indigenous races. There was only one opinion on the policy of England – namely, that the system

138

Plate 6.3 Toliné, who explains that 'They teach me the Bible, and mathematics, and geography' (Verne 1877: 2.154).

tended to the extinction of conquered tribes, and their annihilation in the regions their ancestors inhabited. This fatal tendency is everywhere manifest, and in Australia more strikingly than elsewhere. In the early days of the colony, the convicts, and even the colonists, considered the blacks as wild animals. They hunted and shot them; they massacred them, and sought legal opinion to prove that the Australian black being outside the law of nature, murdering them was no crime. The Sydney newspapers even proposed an expeditious way of getting rid of the Hunter River tribes, nothing less than poisoning them wholesale.

(Verne 1877: 2.188–189)

He continues, narrating British imperialism in a vocabulary of conquest, violence and brutality – a vocabulary of anti-imperial resistance.

The English, at the beginning of their conquests, evidently employed murder as an aid to colonisation. Their cruelties were attrocious. They behaved in Australia as they did in the Indies, where five million Indians have disappeared, and as they did at the Cape, where a population of a million Hottentots has decreased to a hundred thousand. The aboriginal population consequently decimated by bad treatment and drunkenness is gradually disappearing before a homicidal civilisation.

(Verne 1877: 2.189)

In Van Diemen's Land, Paganel informs his companions, the aboriginal population has fallen from five million at the beginning of the century, to several thousand, and is still falling. As elsewhere, the English – not British – are to blame. Neither Glenarvan, nor any of the others, contradicts the geographer. Had they tried, Verne reminds the reader, they would not have had a leg to stand on, as the facts were 'perfectly incontestable' (Verne 1877: 2.189). Anti-imperial sentiments such as this, though present in original translations, were edited out of some British editions. For example, Evans (1965: 29), who informed readers that Verne 'took little interest in politics', removed Paganel's angriest anti-British remarks from his abridgement, *Captain Grant's Children* (Evans 1964a; Evans 1964b).

Paganel's assault upon British imperialism, while often explicit, is also embedded in his geography. He resists British imperialism by subverting geographical knowledge, both imaginative and cartographic. Paganel shows how Toliné's prize-winning geographical knowledge amounts to little more than acceptance of British imperial ideology, which leads to his own submission to the colonists. Maps, never passive descriptions, are vehicles of imperialism. In the Murray Territory, Paganel explains,

The district bears a significant name on the English map, 'Reserve for the blacks'. Thither the aborigines have been brutally driven by the colonists.

(Verne 1877: 2.188)

Maps of the reserves seem to be play some part in imprisoning the Aborigines. Paganel's geography, in the light of his anarchist and anti-imperial sentiments, is ironic. His facts do not add up to his message. As Martin (1985: 179) suggests, the encyclopedic voyages 'promise a final revelation which does not occur'; the factual world is a self-consciously textual construction. His geographical imagination, more than the vast collection of geographical facts it sometimes appears, subverts the geographical imagery and the imperial mappings at the heart of British imperial ideology. It is a site of resistance.

CONCLUSION

The geography of adventure is profoundly ambivalent. Adventures, including imaginary voyages, are populated with ambivalent geographers and ambivalent geographical imagery. On first impressions, the geographers who narrate and occasionally appear in adventure stories are fact-mongers, preoccupied with geographical facts and surrounded by maps and charts. But facts never add up to very much; they are merely the medium of geographical knowledge. In themselves, geographical facts are meaningless, even deceptive. One early seventeenth-century English satirist, commenting on the duplicity of geographical knowledge, named the hero and narrator of *Mundus Alter et Idem* (1605), a geographical travel story, *Mercurius Britannicus*. Translated literally as British Mercury, Hall's narrator is both messenger and deceiver, whose naturalistic geographical descriptions appear to tell the simple truth, but are really the medium of Hall's satirical, political argument. Similarly, Foigny's narrator and Verne's Paganel, ostensibly preoccupied with geographical facts and realistic geographical descriptions, do more than just present geographical knowledge in palatable form. However naively or honestly geographical descriptions may be constructed, however faithfully geographical facts may be gathered, the meaning of a geographical narrative will never be limited to the sum of its facts.

Geographies of adventure, once in cultural circulation, may become sites of resistance. Geographical narratives open up cultural space, ambivalent in itself, which readers then actively enter into. Geographical imagery, compared to more abstract language, becomes easily detached from the intentions of the author, and easily appropriated – unmapped and remapped – according to the needs and interests of readers. For this reason, geographical narratives prove particularly malleable and adaptable, while the spaces they chart are open-ended, fluid and suggestive. Charting ambivalent spaces, suggestive rather than deterministic, adventure stories may be appropriated by conservative, imperial readers, as Verne's extraordinary voyages were in Britain, but for the same reason they may also open up a cultural space to accommodate fluid, creative, critical politics. Foigny, whose political views seem to have been vaguely formulated and muddled, used the medium of geographically realistic adventure to create a space for political reflection. Verne, whatever his intentions, did the same thing. Writers of adventure stories, creating spaces which 'grip the masses' and have the potential to become sites of resistance, engage in what Frederic Jameson (1988) metaphorically refers to as 'cognitive mapping': charting imaginative geographies in the popular imagination, in which political reflection and political action may be possible. The 'cognitive maps' of adventure story writers such as Verne provided space in which readers were able to position and reposition themselves, to negotiate identities and geographies. Since the end of the Second World War, as the British Empire has been dismantled, geographies of adventure have provided space in which readers and writers, in both colonising and colonised countries, have deconstructed colonial and constructed post-colonial identities, as I argue in Chapter 7.

7

UNMAPPING ADVENTURES
Post-colonial Robinsons and Robinsonades

The cultural space created by geographical narratives such as adventure stories, not always confining the imagination to entrenched ways of thinking, to conservative constructions of geography and identity, is a point of departure from which it is possible to deconstruct and reconstruct, unmap and remap geographical and social worlds. To unmap literally is to denaturalise geography, hence to undermine world views that rest upon it. Metaphorically, unmapping means denaturalising more abstract constructs, such as race and gender, which are mapped in imaginative geography. Unmapping is a critical project, a form of resistance to received or mapped world views. Resistance may take the form of critical readings, as I argued in Chapter 6, although it may also take the more tangible form of critical writing. Critical writers appropriate and subvert language and narratives, producing counter-currents of resistance to discursive authority.

Colonial narratives including adventures provide points of departure for criticism and resistance. It was Shakespeare's Caliban who told the European colonists, Prospero and Miranda, 'You taught me language; and my profit on't / Is, I know how to curse' (*The Tempest* I. ii .363–364).[144] Since the native islander, 'savage' and 'slave' to the Europeans, relies upon the English writer to put words into his mouth, and since he is confined to the language of the coloniser, his resistance is modestly preliminary; but resistance it is. Anti-colonial and post-colonial resistance, more generally, may spring from the language of the colonist, from the narratives and imaginative spaces of colonial discourse. Adventure narratives, historically if not intrinsically colonial, provide one medium in which resistance to colonialism is possible. *Robinson Crusoe*, the most famous modern colonial adventure story, has been particularly popular among critics as well as champions of empire. In response to Robinsons and Robinsonades, particularly those published in nineteenth-century Britain, which mapped colonial identities and geographies, a counter-tradition of critical Robinsons and Robinsonades has emerged. Writers from Defoe's contemporary Jonathan Swift (in Britain and Ireland) to post-war novelists including William Golding (Britain), Sam Selvon (Trinidad and Britain), Michel Tournier (France) and J.M. Coetzee (South Africa), have all resisted colonial constructions of identity and geography

143

by writing critical Robinsons and Robinsonades – critical retellings of Defoe's narrative. In *Lord of the Flies* (1954) Golding re-entered the narrative space of Ballantyne's Robinsonade *The Coral Island*, subverting the story and unmapping its vision of British manliness, re-evaluating what it meant to be (white, male and) British in the context of British imperial decline. *Lord of the Flies* goes some way towards decolonising British masculinity. More actively critical – promoting rather than just responding to decolonisation – Selvon, Tournier and Coetzee re-enter the narrative space of *Robinson Crusoe* and, rather than contesting what was constructed within that space, they unmap the space itself, destabilising the terrain on which its particular masculinist, racist, imperialist vision was constructed.

Critical Robinsons and Robinsonades, although sometimes labelled 'anti-adventures', are written within the adventure tradition, and illustrate the ambivalence of adventure's geography. Since critical writers subvert the *Robinson Crusoe* story with deliberate twists and discontinuities that distinguish theirs from more traditional versions, critical Robinsons are noticeably different from the adventure stories that people are most accustomed to. But to classify Robinsons into conservative 'adventures' and critical 'anti-adventures', as some critics have done (Green 1990), is to obscure something of what all adventures – including Robinsons and Robinsonades – are, and to represent adventures as more conservative and politically limited than they are. While the differences between individual adventure stories are often deliberate and important, and should not be glossed over, they are overshadowed by the similarities. These similarities enable some critics to speak of narratives from Defoe's *Robinson Crusoe* and Wyss's *Swiss Family Robinson* to 1950s versions of the Crusoe story including Muriel Spark's *Robinson* (1958) and Luis Buñuel's *Robinson Crusoe* film (1954), as constituents of a single genre (Short 1991; Woods 1995). Each of these stories breaks with some of the traditions and some of the politics of predecessors in their medium. Even *Robinson Crusoe*, the colonial 'straw man' who provides so many post-colonial critics with points of departure, was something of a radical in his day, as I have explained. Indeed, *Robinson Crusoe* broke with so many traditions of adventure that critics and readers argued – and continue to argue – about whether it was an adventure at all. While, for example, Green reads *Robinson Crusoe* as the seminal modern adventure story, Zweig argues that Crusoe 'undermines the ethos of adventure', since he is essentially social and rational, preoccupied with civilised values (Zweig 1974: 113). I suggest, instead, that *Robinson Crusoe* was neither the first nor the last adventure story, but one in a fluid, dynamic and always potentially critical tradition. Similarly, other stories – notably Conrad's *Heart of Darkness* – have sometimes been interpreted as *either* conservative adventures *or* critical anti-adventures, when really they are both adventurous and critical.

Critical adventures, like their more conservative (colonial) counterparts, are capable of reaching broad audiences and making an impression on popular readerships, packaging powerful political and geographical messages in an appealing, readable narrative form. Criticism, in the form of adventures and other readable narratives, is more accessible and therefore potentially more effective than it would be in less popular works such as academic monologues and experimental literature. Academic critics and avant-garde novelists may subvert adventure stories and comment on adventure stories, but they generally speak to relatively small audiences, sometimes just to each other. Among the first critics of *Robinson Crusoe*, for example, was Defoe himself. Defoe commented on his own adventure story in the second and third volumes of *Robinson Crusoe*. In volume two, *The Farther Adventures of Robinson Crusoe* (1719b), the island utopia, ceded to a party of mutineers, falls into a dystopian state. In volume three, *Serious Reflections During the Life and Surprising Adventures of Robinson Crusoe* (1720), Defoe reflected upon his realistic style and qualified his own truth claims. He was so self-critical that he would probably have undermined his own story had he succeeded in engaging the general reading public as he had in volume one. But as he ceased to tell a story, and began to comment on it, Defoe lost his readers' attention and, at the same time, he lost his power to map and unmap. Other writers fail to unmap popular geographies because they, too, stray too far from the traditional narrative forms of adventure, and from lively and accessible story-telling more generally. Robert Kroetsch, for example, commented on western Canadian quest adventures in his clever, esoteric novel *Badlands* (1975).[145] In Kroetsch's story a questor's daughter unwrites and unmaps his quest by retracing his steps, drunkenly shredding his journal and undermining its authority. But Kroetsch's critically acclaimed novel made a bigger impression on students of Canadian literature than consumers of popular culture. Similarly, academic criticisms of and commentaries on adventure are often read under coercion, when on college reading lists for example, and as a result they lack the power to influence popular imaginations. Even George Orwell, among the more readable of the 'serious' critics who have turned their attentions to adventure literature, failed to make much of an impression on popular attitudes to adventure in his wartime assault upon boys' weeklies. Peter Hunt, the Welsh children's story writer and critic, has suggested that Orwell's failure to bury the right-wing boys' weeklies was due, in part, to his humourless, quasi-academic tone.[146] Academic critics sometimes accuse adventure writers of seducing readers, checking their critical distance and moral judgement by drawing them into romance, fantasy and adventure narratives (for example, Katz 1987). Although these seductive narrative techniques are not the only ones that engage readers, they do suggest a lesson for academic and other critics: to be effective, to be listened to and/or read, one must be able to tell a good story. Critical adventure stories, like all adventure stories, are first and foremost stories; the story carries the criticism rather than vice versa. The critical adventures that make the biggest impression on popular audiences, and therefore map or unmap most effectively, are the ones that tell a story that is gripping enough to hold the attention of readers, who may or may not be aware of or interested in the author's specific literary and political references. Thus, while Orwell condemned the 'Conservative bias' of boys' weeklies, he did not condemn those weeklies outright. Inspired by Spanish 'anarchist' and 'left-wing novelettes', he suggested that British boys' weeklies might be turned to the left, that their adventures might be appropriated by Socialists (Orwell 1940: 126). Post-colonial critics have accepted Orwell's challenge, turning adventure around, if not to Orwellian Socialism. In novels such as *Lord of the Flies* and *Friday*, for example, writers such as Golding and Tournier respectively show how one way of negotiating what an adventure story has mapped, one way of questioning its values and denaturalising its world view, is to write another adventure story.

Despite adventure's historical association with colonialism, or rather because of it, adventure has sometimes been a vehicle of post-colonial resistance, the geography of adventure a site of resistance. This is true, perhaps more than ever, of the post-war period. Some post-war writers have tried to remap identities and geographies, in the context of crumbling colonial empires, while others have struggled more actively against those empires and their legacies. As Europe's colonial empires dissolved, political geography changed fast. Maps, of identity as well as geography, quickly went out of date. John Strachey's *The End of Empire*, published as late as 1959, could be marketed (on the dust jacket) as

the first book to face up to the situation in which Britain finds herself as the result of the voluntary dissolution of her Empire during the past fourteen years. Its object is not only to explain what has happened and why, but to show how great are the opportunities for Britain in 'her post-imperial period'.

As Strachey (1959: 140) wrote, decolonisation, 'still only imperfectly realised by the British people', was dramatic and dizzyingly fast. In 1945, 550 million of the world's 2,225 million inhabitants had been ruled from Whitehall, while another fifty million lived in 'semi-colonies' such as Egypt and the Arab states. By 1959, over five hundred million of those six hundred million people were completely self-governing. Most of the remainder – approximately eighty million people – were on their way to self-government. 'Moreover,' Strachey reminded the reader, 'this process of dissolution continues uninterruptedly' (Strachey 1959: 140). At the very least, decolonisation presented the opportunity or necessity for Europeans, particularly Britons, to rethink their place in the world, hence to rethink their identities. Strachey cautiously welcomed the demise of 'that demonic will to conquer, to rule, and sometimes to exploit' (Strachey 1959: 217). Golding, in *Lord of the Flies*, sought to reposition himself in a decolonising world, to decolonise metropolitan masculinity. Others, less content to *observe* the demise of empires, and more sensitive to the forms of imperialism that live(d) on, actively pursued and pursue the ongoing project of decolonisation. Said argues that

The slow and often bitterly disputed recovery of geographical territory which is at the heart of decolonisation is preceded – as empire had been –

by the charting of cultural territory. After the period of 'primary resistance', literally fighting against outside intrusion, there comes the period of secondary, that is, ideological resistance.

(Said 1993: 252)

Post-colonial artists and intellectuals, seeking most generally to 'decolonise the mind' (Thiong'o 1986), unmap colonial geographies from the material and metaphorical margins, 'challeng[ing] the impact of imperialism on non-Western cultures' (Jackson 1995: 465).[147] 'Writing back to the centre' (Ashcroft, Griffiths and Tiffin 1989), adventure writers including Selvon, Coetzee and Tournier create spaces of post-colonial resistance.

UNMAPPING IMPERIAL MASCULINITY: RETURNING TO BALLANTYNE'S ISLAND

William Golding returned to the island setting of Ballantyne's Robinsonade *The Coral Island*, subverting the original story and unmapping the British manliness it naturalised.

In the early 1950s, with Britain in post-war ruins and the British Empire in decline, the geography of adventure stories like *The Coral Island* was out of date, remaining as the hangover of a previous era, a persistent but anachronistic literary map. William Golding, a teacher in an English boys' school, knew that Victorian adventure stories like *Treasure Island* and *The Coral Island* were still favourites among British children, as they had been among their parents, particularly their fathers (Turner 1957) (Plate 7.1).[148] Orwell's opposition to boys' weeklies sprang from his awareness of their influence on young (and old) minds. Even Winston Churchill acknowledged that adventure stories framed and coloured his 'real' experiences (A. White 1993: 59). Stevenson's and Ballantyne's Robinsonades never bore as close a relationship to 'reality' as to geographical fantasy and imperial ambition, but their relationship to British civilisation and imperialism became very tenuous after the war. The unbounded confidence and optimism of Ballantyne, plausible in Victorian Britain, at least among juvenile readers, was not at all plausible in the early 1950s. Ballantyne's confidence in human nature, or rather in the ultimate decency and superiority of white, European, Christian men, was undermined by the experience of war, the evidence of what such men are capable of doing to each other. And Ballantyne's optimism with regard to the progress of British civilisation was contradicted by the state of British cities and industry, which were devastated by war, and the state of the British Empire, which was in rapid decline. Britain no longer seemed like the geographical and historical centre of the world, as it had been depicted in stories like *The Coral Island*. As Strachey (1959: 217) suggested, Britain needed 'some fresh national purpose, capable of inspiring the spirit and energies of the British people', or at least a fresh sense of national identity. Golding took this disjunction between the world he encountered in *The Coral Island* and the

147

THE
CORAL ISLAND
by
R. M. Ballantyne

*Illustrated with line drawings
and 8 colour plates by*
LEO BATES

LONDON: J. M. DENT & SONS LTD
NEW YORK: E. P. DUTTON & CO. INC.

Plate 7.1 Frontispiece and title page of *The Coral Island*, published by J.M. Dent (London) in 1957. A gaudily illustrated mid-twentieth-century edition, contemporaneous with early editions of *Lord of the Flies*.

world he experienced as a British naval officer in the war and as a teacher in post-war Britain as his point of departure for writing *Lord of the Flies*. Golding, seeing 'Ballantyne's book as a badly falsified map of reality, yet the only map of this particular reality that many of us have' (Niemeyer 1961: 243), set about unmapping Ballantyne's geography.

Lord of the Flies is a re-enactment of *The Coral Island*, with a similar cast of characters, a similar initial scenario and a similar setting. Golding takes Ballantyne's original cast of boys and adds more. Ralph, Jack and Peterkin in Ballantyne's story become Ralph, Jack, Piggy and Simon in Golding's. Golding also mentions at least fourteen others by name, although there are more, perhaps as many as sixty, in his story. The boys are plane-wrecked (the modern equivalent of shipwrecked) on a vaguely located desert island that seems (at first) like paradise, and is (probably) located somewhere *en route* to Australia (Oldsley and Weintraub 1965). Golding's story begins in much the same way as Ballantyne's, as the boys realise that they are alone on an island, where they may play, have adventures and find some way of surviving, all in the absence of adults.[149]

Golding's boys find themselves on an island much like the one in *The Coral Island*, and they see themselves and their island through the eyes of adventure writers like Ballantyne. After Ralph, Jack and Simon explored, and found themselves to be on 'a good island' with 'food and drink', with outcrops of pink

148

granite and patches of blue flowers, they become optimistic about their adventure (Golding 1954: 45). The island seems to spring from the pages of an adventure story. 'Here at last was the imagined but never fully realised place leaping into real life' (Golding 1954: 21). And the boys compare themselves to the heroes of adventure stories they have read.

'It's like in a book.'
At once there was a clamour.
'Treasure Island – '
'Swallows and Amazons – '
'Coral Island – '
        (Golding 1954: 45)

The Coral Island is not the only adventure the boys know and it is not the only adventure story Golding refers to.[150] Compared with other stories, such as *Treasure Island* and *Swallows and Amazons* (1930), *The Coral Island* is particularly overt in its masculinism and imperialism. Less a representative than a caricature of adventure, it is among the least subtle in its portrayal of Christian manliness and British imperialism. *Treasure Island* is set principally on an exotic but festering island, not in itself an object of colonial desire, not populated with 'savage' enemies (the whites are savage enough), and not visited by muscular Christians. And in *Swallows and Amazons*, Arthur Ransome exchanged realistically exotic settings for geographical fantasies, for imaginary encounters with Amazonian pirates in England's Lake District (see Hardyment 1984). It is Ballantyne's exuberant adventure that provides Golding with his clearest point of departure, and with his setting.

Golding's story soon diverges from Ballantyne's, but the setting continues to function in basically the same way. The island setting of *Lord of the Flies*, like *The Coral Island*, *Gulliver's Travels*, *Robinson Crusoe* and a host of other modern British adventure stories, is a vague, malleable, simplified space, a microcosm or caricature of contemporary Britain. As Hannabuss (1983) put it, these islands were 'laboratories' in which to isolate and examine aspects of Britain. While Defoe's island was a space in which to visualise British Christian *petit bourgeois* society, and Ballantyne's was a space in which to imagine a form of British manliness, Golding's is a space in which to re-examine the same boys, as specimens of British manliness and as human beings. Golding, like Defoe and Ballantyne, describes the island in naturalistic detail. But, also like Defoe and Ballantyne, he is not really interested in geography or natural history, and his island remains geographically vague and inconsistent (Oldsley and Weintraub 1965). Naturalistic detail serves to naturalise the ideas – about society and humanity – that are presented in the story. Golding's island is defined mainly by its 'appearance of reality' (to borrow Defoe's phrase) and its disconnectedness and remoteness from Britain. Although superficially realistic, it, like Defoe's and Ballantyne's islands, is mainly an imaginative space in which to reinvent Britain and British manliness. In a setting much like Ballantyne's, Golding subverts Ballantyne's story and, in doing so, subverts the visions of Britain and British manliness his predecessor mapped.

Both Ballantyne and Golding define the island by its geographical remoteness from Britain, but whereas Ballantyne represents it as a frontier of ever-expanding British civilisation, Golding represents it as a refuge from the smouldering remains of British or, more generally, western civilisation. Golding's story begins as boys are being evacuated from Britain, some of them vaguely aware that an atom bomb has hit England. 'Didn't you hear what the pilot said? About the atom bomb?' one boy repeats, 'They're all dead' (Golding 1954: 20).[151] Golding's point of departure is the end of civilisation as the British have known it, the end of history. As a reminder of the state of the civilised, war-torn adult world, a dead parachutist floats down from the upper air, and becomes entangled in a tree. Golding later explained that the dead parachutist was meant to represent History – a western master narrative, recently expired. 'All that we can give our children' is the hideous, decaying adult, who is 'dead, but won't lie down' (Golding quoted by Kermode 1961: 19). In *Lord of the Flies*, as in *The Coral Island*, the boys confront 'savagery' on the island, but while Ballantyne's boys succeed in defeating it with their 'civilised' ways, Golding's find that it is their 'civilisation' that is defeated. Ballantyne's boys are always fun, manly, civilised, complimentary, loyal, friends. Golding's boys begin with rules and orderly meetings, wearing clothes and keeping a fire burning, but they gradually lose their grip on these symbols of civilisation. They become obsessed with fears of a 'beast', and revert to a savage condition, exchanging their clothes for paint, and spilling the blood of pigs and, later on, each other (they kill Simon then Piggy). Whereas Ballantyne's boys confront savagery and evil among the cannibals and pirates they meet in the South Pacific, Golding's boys find savagery and evil among themselves, as a group and as individuals. Whereas Ballantyne's boys confront tangible geographies of darkness, Golding's confront the darkness within themselves, 'the darkness of man's heart' (Golding 1954: 248). Like Conrad's Charlie Marlow (in *Heart of Darkness*) who, as a 'little chap', stared at maps and dreamed the seemingly innocent dreams of boyhood adventure, Golding's boys embark upon the adventure they have dreamed of, but in so doing they find out things, mainly about themselves, that frighten them.

The boys in Golding's story are very much aware how far they have come from the adventure they once seemed to be embarking upon, the adventure narrated in the books they had read in England. Ralph's memory of boys' adventure stories (he mentions that he never read girls' stories) is juxtaposed with his own adventure. He remembers books at home, which

> stood on the shelf by the bed, leaning together with always two or three laid flat on top because he had not bothered to put them back properly. They were dog-eared and scratched. There was the bright, shining one about Topsy and Mopsy that he had never read because it was about two girls; . . . there was the *Boy's Book of Trains*, *The Boy's Book of Ships*. Vividly

they came before him; he could have reached up and touched them, could feel the weight and slow slide with which the *Mammoth Book for Boys* would come out and slither down. . . . Everything was all right; everything was good-humoured and friendly.

The bushes crashed ahead of them. Boys flung themselves wildly from the pig track and scrabbled in the creepers, screaming.

(Golding 1954: 139–140)

Near the end of Golding's story, as Ralph is being hunted down by Jack and the other boys, a British naval officer appears on the beach. The officer, learning that the boys have been plane-wrecked on the tropical island, remarks 'Jolly good show. Like the Coral Island' (Golding 1954: 248). Ralph knows only too well that it has been nothing like *The Coral Island*. In a final, tearful scene, he and his companions understand how far they have come from Ballantyne's idealised, seemingly innocent childhood adventure, and how different they are from Ballantyne's pure, manly heroes.

Like Ballantyne almost exactly a century before him,[152] Golding responded to and articulated a historical, geographical and intellectual moment. At the end of the decade (the 1950s), *Lord of the Flies* could be called (by a prominent literary critic) 'the most important novel to be published . . . in the 1950s' (C.B. Cox 1960: 112). Just a few years later another critic, re-reading *Lord of the Flies*, commented that its 'pessimism and spiritual fatigue' seemed 'strangely dated' (Grande 1963: 457). The appeal of Golding's story was by no means limited to Britain in the 1950s, but its most immediate appeal and relevance was in that context.[153] I do not wish to dismiss Golding's more universal references, nor to deny the importance of other situated readings of his story, but I am interested in *Lord of the Flies* as a story written and received within the specific historical and geographical context of post-war Britain. While Ballantyne asserted a world view – including a vision of British manliness – that seemed plausible to mid-nineteenth-century Britons, Golding did the same for his own particular, war-torn and disillusioned generation.[154]

Lord of the Flies illustrates the potential of adventure stories to unmap as well as to map, to challenge as well as to assert. Critics have argued about how to classify Golding's story, whether to call it a novel or a fable, for example. I will call it an adventure story.[155] While Golding turned Ballantyne's story on its head, he still told a gripping Christian[156] tale of white male adventure on a tropical island, and he still presented an apparently simple world view (in some respects the opposite of Ballantyne's) with absolute, manly confidence (Billington 1993). Golding unmapped from roughly the same position as Ballantyne mapped, that is, as a white, heterosexual, middle-class, literate, Christian, British man, who wrote and published in Britain, and who was well-received by readers in Britain and overseas. In this sense, at least, he was a very similar kind of adventure story writer to Ballantyne. Both Golding and Ballantyne used adventure as a medium in which to negotiate British manliness, imagining it in and in relation to exotic settings of adventure. In the adventure narrative, and in its imaginative geography, it is difficult or impossible to get very far from imperialism. Said, in his discussion of Conrad, argues that post-colonial resistance cannot get very far so long as it is confined to the cultural terrain of colonialism. To fundamentally unmap colonial geographies and identities, it is necessary to destabilise the terrain upon which colonialism is normalised and naturalised.

UNMAPPING DEFOE'S ISLAND: DENATURALISING CRUSOE'S WORLD

To unmap fundamentally the world view of adventurers like Ralph Rover and Robinson Crusoe, it is necessary not only to re-enter and subvert their narratives, but also to contest the terms – the language and the geography – in which those narratives are constructed. It is necessary to go further than Golding did when he challenged Ballantyne's world view – on his predecessor's terms and in his territory. But, in so doing, it is not necessary to abandon adventure; it is possible to unmap particular adventures by writing new, critical adventure stories. Post-colonial writers, retelling *Robinson Crusoe*, begin where Golding left off. They subvert the adventure story that played such a strong role in naturalising and legitimating a British imperial world view, not only in the minds of the colonisers but also of the colonised,[157] first by turning the tables on the white, male British coloniser, then by undermining his language and his setting. Selvon, Tournier and Coetzee re-enter the narrative space of *Robinson Crusoe* and begin to retell the story from the perspective of a woman (absent in Defoe's story) and a black man (Friday, colonised in Defoe's story). But rather than appropriating Defoe's story and contesting his world view on his own terrain, they unmap Defoe's geography, destabilising the ground on which its masculinist, racist, imperialist vision was constructed. Although they use *Robinson Crusoe* as a point of departure for their post-colonial projects, they retain some elements of *Robinson Crusoe*, including its general geographic structure.

Post-war, post-colonial subversions of *Robinson Crusoe* that appropriate, undermine and unground that story are pre-dated by many earlier Robinsonades, notably Jonathan Swift's *Gulliver's Travels* (1726).[158] Jonathan Swift, a contemporary of Defoe, opposed British imperial expansion, and was particularly critical of colonial trade (including slavery) and annexation, which Defoe promoted (or at least tolerated). He subverted the geography of *Robinson Crusoe* – and therefore the colonial encounters it naturalised, promoted and legitimated – in *Gulliver's Travels*.[159] Gulliver, a voluntary castaway, resolves to re-enact the *Robinson Crusoe* story.

> My Design was, if possible, to discover some small island uninhabited yet sufficient by my Labour to furnish me with the Necessaries of Life, which I would have thought a greater Happiness than to be first Minister in the Politest Court of Europe; so horrible was the Idea I conceived of returning

to live in the Society and under the Government of *Yahoos*. For in such Solitude as I desired, I could at least enjoy my own Thoughts, and reflect with Delight on the Virtues of those inimitable *Houyhnhnms*, without any Opportunity of degenerating into the Vices and Corruptions of my own Species.

(Swift 1726: 4.165–166)

But soon after setting foot on such an island, Gulliver is wounded in the knee with the arrow of a hostile native. He escapes in despair on a Portuguese trading vessel. Thus Swift summarily dismissed Defoe's island story. More generally, Swift subverted Defoe's realistic geography. He satirised Defoe's realistic style, the style of a travel writer seeking to convince his readers he was telling the truth, with detailed but nonsensical descriptions of travel. During a storm at sea, for example, 'We got the Star-board Tacks aboard; we cast off our Weather-braces and Lifts; we set in the Lee-braces, and hawl'd forward by the Weather-bowlings' (Swift 1726: 2.4).[160] The geography of *Gulliver's Travels* is realistic in appearance, finely detailed, but impossible. For example, Swift located the islands of Lilliput, Blefuscu and Houyhnhnm-Land within the shores of New Holland, not by mistake or misprint (as some critics have suggested) (Moore 1941), but as part of his burlesque of travel literature (Gibson 1984). Undermining Defoe's geography, Swift undermined the imperial encounters it glorified and naturalised. His strategy of denaturalising Robinson Crusoe's colonial encounter by subverting his story and his geography is echoed in critical, post-colonial retellings of the story. Whereas Swift, a privileged and literary member of the British[161] establishment, retold *Robinson Crusoe* from a (critical) perspective close to the 'centre', others retell it from the margins, material and metaphorical. So whereas Swift was anti-colonial, some others can more properly be described as post-colonial. To 'write back to the centre', as post-colonial writers do, is to write from the margins of empire, contesting colonial versions of geography and history (Ashcroft, Griffiths and Tiffin 1989, 1995; Crush 1994). Writers such as Selvon, Tournier and Coetzee are in some senses marginal figures, although their biographies, like those of most other people, combine elements of privilege with elements of marginalisation. Their works are illustrative rather than representative of post-colonial literature that 'writes back to the centre'.

Post-colonial Robinsons begin by appropriating the narrative of the white, male, British coloniser. Sam Selvon reverses Defoe's order of white and black, master and slave, coloniser and colonised, England and island, in *Moses Ascending* (1975). Unlike Coetzee and Tournier, Selvon writes from the (approximate) perspective of his narrator and protagonist – a black, Trinidadian man living in London. He re-enters the narrative space of *Robinson Crusoe*,[162] maintaining its realistic appearance and its generally binary spatial structure, but reversing the roles of England and the tropical island and, with them, the roles of white and black people. Moses, the black narrator and Crusoe figure, devotes most of his attention to the basic problems of survival in London. He finds a precarious shelter in the form of a house, condemned by the City of London, which he rents out to tenants. He is aware that he is acting out some kind of Robinsonade. Away from home – the 'sandy beaches and waving coconut palms' of the West Indies (Selvon 1975: 117) – in an uncomfortable, lonely, foreign place he struggles to survive. In the 'penthouse' of his condemned dwelling he eventually finds the time and quiet space he needs to begin writing his memoirs. The struggle to survive is the basis for his story, and leads to the few adventures, reluctantly embarked upon, which he describes. These arise when Moses gets mixed up with some Black Panthers from America, and when he assists illegal immigrants from Asia. To immigrants from Asia and the West Indies, London is a strange land. The immigrants arrive in London much as Crusoe, with goods salvaged from his ship, arrived on the island.

> It was a motley trio that Faizull shepherd into the house. I have seen bewitched, bothered and bewildered adventurers land in Waterloo from the Caribbean with all their incongruous paraphernalia and myriad expressions of amazement and shock, but this Asian threesome beat them hands down.
>
> (Selvon 1975: 74)

Moses, the Black Crusoe, finds white natives in England. He 'saved' a man Friday, an illiterate, non-Christian Englishman from the *Black* Country, giving him a home in exchange for his labour. 'Witness how I take in poor Bob, and make him my footman, when he was destitute and had no place to go when he land in London' (Selvon 1975: 32). Blacks convert whites to Christianity rather than vice versa. A Black Panther, attempting to 'spread the gospel to the white heathens', feels that 'not enough black missionaries like himself infiltrated the white jungles' (Selvon 1975: 100). Blacks teach whites to speak and write English. Galahad, a black acquaintance of Moses, helps Bob with his pronunciation; 'you know what an awful accent these Northerners have' (Selvon 1975: 147). Moses tries to civilise Bob. 'As we became good friends, or rather Master and Servant, I try to convert him from the evils of alcohol', and 'decided to teach him the Bible when I could make the time' (Selvon 1975: 11).

Selvon parodies *Robinson Crusoe*, unwriting a colonial story and unmapping a colonial world while writing a post-colonial story and mapping a particular post-colonial world view. But Selvon does not do this entirely within the terms – the language, the narrative and the geographical structure – defined by Defoe. While he begins by reversing the roles of Crusoe and Friday, England and the island, he proceeds to dismantle the order of *Robinson Crusoe* more completely, to deconstruct its racially and geographically dualistic world view. Having switched the roles of Crusoe and Friday, white and black, he switches them again. The white Friday rises up against his black master. Near the end of *Moses Ascending* Moses loses his penthouse to his man Friday, Bob, who then demands to be called Robert (almost 'Robinson'?). Moses has to move into Galahad's uncomfortable basement. Roles and spaces turn out to be fluid, interchangeable, or, perhaps, the old white/black master/slave relationship is reasserted. The authority of the English language, too, is undermined. Moses lives in 'Brit'n', not Britain. Speaking and writing West Indian 'english', and teaching it to Bob, Moses refuses to defer to the authority of British (Received Standard) 'English'. Whereas Defoe's Crusoe spoke in standard (if journalistic) English, contrasted to Friday's perpetually broken English, Moses speaks in an 'english' reminiscent of Friday's, which is contrasted to Bob's 'awful' English. Beginning to dissolve the dualistic identities and spaces of *Robinson Crusoe*, and to undermine the language in which it was written, Selvon attacks some of the foundations of *Robinson Crusoe*. But the main force of *Moses Ascending* comes from appropriation – rather than deconstruction – of the white, British colonist's story.

To switch narrators, and thus retell and remap *Robinson Crusoe* from a different point of view, is ultimately to remain within the same narrative space. Selvon transposes the story and the setting of *Robinson Crusoe* almost beyond recognition, but he retells the same essential story (from the perspective of Friday) and remains within the same (binary) geographic structures. To challenge *Robinson Crusoe* even more fundamentally, it is ultimately necessary to step outside of his narrative space, to deconstruct his geography.

Michel Tournier steps outside Defoe's setting – outside Robinson Crusoe's realistic island – by exploring a number of points of view but exposing each as a textual construct. Like Selvon, Tournier begins to rewrite *Robinson Crusoe* from the perspective of Friday then proceeds to subvert it. But Tournier's island setting and black inhabitant, never realistic, are overtly the constructs of European literature, not 'real' colonised places and peoples.

The title of Tournier's version of the *Robinson Crusoe* story – *Vendredi, ou les Limbes du Pacifique*, translated as *Friday; or The Other Island* (1969), and followed by a children's edition, *Friday and Robinson; Life on Speranza Island* (1972)[163] – suggests that Tournier's intention is to retell the story from Friday's perspective. But the story is told by Robinson, not Friday. Tournier does not introduce Friday until the second half of the story. By then Robinson has already been shipwrecked, reverted to a state of near insanity – wallowing in the mud and hallucinating – and then restored his 'civilised' ways, with regular work, rules and a residence. Nevertheless, when Friday does finally arrive he gradually assumes centre stage in Robinson's story. At first Friday plays the part of Robinson's slave. But he never takes his master's rules or regimented work practices very seriously. He never quite gives in to his master's demands, never renounces his freedom; he is always 'natural' man. From Friday's perspective, as a detached, bemused onlooker, the 'civilisation' Robinson takes so seriously – his superficially practical British colony – appears ridiculous. Friday's careless attitude leads him to make mistakes, to ruin a field of his master's crops and, ultimately, to destroy Robinson's stores and residence by igniting the gunpowder. Robinson, rather than angry, is relieved to lose the civilisation that was, in effect, his cage. At this point Friday becomes his mentor, teaching him how to make music and good food. In the pen-and-ink illustration, a 'fantastic assembly of lords and ladies' (Plate 7.2), Friday has dressed the cactus plants in Robinson's best clothes and hats, and is dancing naked among them, in a gesture of good-natured, playful disrespect towards Robinson's 'civilisation'. The image, reminiscent of illustrations by the Victorian aesthete Aubrey Beardsley,[164] lacks detail and any sense of reality, and is decadent and passionate, evoking sensuous pleasures of the body and the earth. In this heady atmosphere, Robinson soon follows Friday on the path towards nature, and becomes a 'natural' man, suntanned and naked. Friday's way of living, and his way of seeing the island, have supplanted Robinson's.

Although Friday assumes centre stage, and defines his own stage, he remains a term in Robinson's story, never a non-European or non-white speaking for himself. Specifically, Friday is a predefined literary figure, an image from Romantic literature. Tournier signals his Romantic vision of Friday explicitly, setting his story exactly a century after the original. Robinson is a contemporary of Rousseau, and his shipwreck coincides with the writing of *Emile* (Purdy 1984), the work in which Rousseau recommended *Robinson Crusoe* to young readers and identified Crusoe as 'natural man' (not what Defoe intended).[165] Friday is mainly an image in Robinson's Romantic imagination, which had begun to develop before Friday arrived. Before he encountered Friday, Robinson was in the habit of retreating from his 'civilised' spaces and ways, crawling into a dark, womb-like cave and making symbolic love to the earth. Friday does not give him the idea of retreating from society to nature, he just shows him how, and shows him that in doing so he need not revert to mud baths and hallucinations.

Friday helps Robinson to see a different island, but rather than showing him the 'real' island, he shows that no island is real. Tournier does not try to rewrite *Robinson Crusoe* and remap the island from the perspective of a colonised, non-white, non-European. *Friday*, described by one reviewer as 'heady French wine in the old English bottle' (Fleming 1969), is more a parody of such a project, particularly as it might be handled by a white, French, literary man. Tournier's island is never allowed to look much like reality – to have the 'appearance of reality' – for very long. It is an explicitly literary construct, malleable and insubstantial. Reviewing the book, Jonathan Raban suggests that 'Robinson's sturdy capacity to create the world in which he lives is as brave as it is comic' (Raban 1969: 217). The island is unstable – anything and nothing. As the epigraph (attributed to Jean Guéhenno) puts it, 'There is always another island'. On this shifting ground, world views and imperial encounters flash before the reader's eyes, but they never remain still long enough to seem real.

To go further than Tournier, who steps outside of Robinson Crusoe's island, would be to step outside his story, to break down his narrative. This may mean assuming critical distance from adventure, subverting adventure rather than writing critical adventures. Booker prize-winning novelist J.M. Coetzee embarks upon such a project in *Foe*, his retelling of the *Robinson Crusoe* story. *Foe*, like *Moses Ascending* and *Friday*, begins as a retelling of Robinson Crusoe from the

Plate 7.2 Illustration of a 'fantastic assembly of lords and ladies', hosted by Friday, in *Friday and Robinson* (1972).

perspective of a marginalised character, in this case a woman, absent in Defoe's story and in most Robinsonades. Given Coetzee's preoccupation – as a white, South-African enemy of apartheid (Penner 1989) – with race, it is not surprising that *Foe* develops not as a (white) woman's appropriation of a (white) man's story, but as a third-person perspective on the relationship between the white man and his black slave, and then as a more general subversion of the story, in which Coetzee undermines the project of appropriating it from any perspective.[166]

In *Foe*, Coetzee assumes the perspective of first-person narrator Susan Barton, who is cast adrift in a rowing boat.[167] Barton had been left to fend for herself after mutineers killed the captain of her ship and raped her.[168] After rowing and then swimming to a nearby island, she is taken in by Cruso (Coetzee's spelling and, occasionally, Defoe's) who lives there with his slave, Friday. Barton does not find the hero described in Defoe's story, but a forgetful old man with rotting teeth. Friday, too, is dull, mute and generally less romantic than Defoe's Friday. The island is not quite as it appeared in Defoe's tale, or as desert islands appeared in travellers' tales more generally. Coetzee spells this out, through the voice of Barton.

> For readers reared on travellers' tales, the words *desert isle* may conjure up a place of soft sands and shady trees where brooks run to quench the castaway's thirst and ripe fruit falls into his hand, where no more is asked of him than to drowse the days away till a ship calls to fetch him home. But the island on which I was cast away was quite another place.
>
> (Coetzee 1986: 7)

When Barton cross-examines Cruso on the subject of his past life, shipwreck, island life and sole companion, Friday (whose tongue seems to have been cut out), she finds his stories inconsistent and confused, and doubts both his memory and his sanity. Cruso lives in his own private world, not observing the island or communicating with those around him. He has taught Friday just enough that he will obey orders, and he does not listen to Barton. He has kept no journal, and he has no idea how long he has been on the island. Barton, Cruso and Friday are later rescued from their island by a passing ship. On the way back to Britain Barton tells her story to the captain, who says that there has never been a female Crusoe and suggests she set her experiences down in writing. Back in Britain, Barton brings her story to a writer named Foe (Defoe's family name), who is to turn it into a book. Cruso, no longer the believable narrator of his own story about himself, his slave and his island, is to become a character in Barton's story. Friday and the island, too, are to fall into Barton's quest narrative, the adventure story of her travels in search for a lost daughter. Barton and Foe, it seems, have a different, more truthful version of the island and its inhabitants, than that told by Crusoe and/or Defoe.

Having re-entered the narrative space of *Robinson Crusoe*, Coetzee then subverts it, undermining its realistic geographical images and denaturalising its world view by destabilising its language and ungrounding its realistic terrain. At first Susan Barton is physically present, a woman whose experiences on the island, swimming in the ocean and recovering on the beach, defecating in the garden and having sex with Cruso, are vivid and very 'real'. In the first third of the story, when she is on the island, Barton's presence is stronger as a character than a narrator, although she is always both. Her narration, like Defoe's in *Robinson Crusoe*, seems to simply report her adventure. She insists on resisting the temptation to describe strange and fanciful fruits, serpents, lions and cannibals. But when she is rescued and returned to Britain, attention is refocused upon her role as a narrator rather than as a character. At first she believes in and insists on telling the truth. 'I will not have any lies told', she insists (Coetzee 1986: 40). But in Britain she has plenty of time to reflect upon, and begin to doubt, the possibility of truth in storytelling. As a narrator, she becomes a more abstract voice, a vehicle for Coetzee to comment and speculate on questions of narrative, truth, language, authorship, representation and colonialism. She also becomes a self-effacing mediator between Cruso and Friday – the characters of her story

– and the binary worlds they represent: master and slave, white and black, articulate and mute, coloniser and colonised.[169] Barton's reflections on language and storytelling and her relationship with Foe, and with a woman who claims to be her daughter, lead her to doubt the veracity not only of her story, but also of herself, her own existence. The realism of the first third of the story is broken down, qualified and undermined, until nothing seems real any more, except possibly the text. Barton asks Foe, 'Am I a phantom too?' (Coetzee 1986: 102). She complains that the island, too, must be part of a story, unable to stand as a thing in itself. Barton realises that the island is just the setting of a story, in which she, Friday and Cruso are the characters; all are textual constructs. Thus Barton qualifies and ultimately abandons her own faith in the essential truth of her realistic images.

Critics have puzzled over why Coetzee seemed to embark upon the project of retelling Robinson Crusoe from a woman's point of view, but then allowed his story to degenerate into a series of general reflections. George Packer (1987: 404) suggests that 'Foe reads as if Coetzee started out to reinvent Defoe's famous tale through a woman's eyes, became intrigued with the linguistic and philosophical implications, and ended up writing a commentary on the elusiveness of his own project.' Some critics see this a failure on Coetzee's part, suspecting that he was side-tracked. Dick Penner (1989: 127–128) has trouble seeing the adventure and the subsequent reflections as part of an 'artistic whole'. He 'wishes that [Coetzee] had devised something more engaging for Barton to do after leaving the island', and complains that too little happens after the return to England. Other critics complain that Foe 'never quite comes to life' (Auerbach 1987: 37) and suggest that 'theories about fiction are best suggested implicitly, rather than through direct discourse' (Packer 1987: 404).

Coetzee does not succeed in writing a critical adventure so much as in criticising an adventure. He does not just rewrite Robinson Crusoe, he rewrites Defoe's whole trilogy. The island adventure ends one-third of the way into the book, just as it ended one-third of the way into Defoe's trilogy. The two sequels to Robinson Crusoe were devoted mainly to reflections on the relation of fiction to reality, and to qualifications of the realism in the original story. Similarly, Foe degenerates into a series of rambling reflections on the subject of truth in fiction, which lead to no particular conclusions, other than a general abandonment of the narrator's faith in realism. Coetzee, although he seems to be informed by late twentieth-century post-structural philosophers, comes to a similar set of conclusions to his eighteenth-century predecessor.[170] But, like Defoe, he strays too far from adventure, by that point, to really subvert (rather than comment on) adventure, and perhaps to hold the interest of some of his more general readers. As it says on the dust jacket of Foe, 'Foe is the didactic, philosophical novel at its most sophisticated, subtle and superb' (1986 Viking edition, quoting Richard Burns, The Literary Review). Praise such as this is sure to frighten off all but the most pious of intellectuals, leaving popular mental maps intact.

CONCLUSION

Golding, Selvon, Tournier and Coetzee illustrate ways in which adventures can be used critically to negotiate world views, constructions of identity and geography. Golding stayed close to the form of the adventure story he subverted, while Selvon, Tournier and particularly Coetzee departed further from it. Coetzee subverted so much of the original adventure that his story strays outside the boundaries of adventure, observing adventure at a distance. But, to the extent that they remain within the bounds of adventure, and resist the temptation to assume critical distance from it, these variously post-colonial writers illustrate ways in which adventure can unmap, leading readers into disturbingly, actively 'unknown' spaces, in which it is possible to invent new identities and geographies.

To unmap Robinsons, Robinsonades and other adventure stories is to open space in which to invent new worlds, to make room for new voices and new constructions of geography and forms of identity. Critical adventures are not just negative exercises in erasing established constructions of geography and identity. They use the geography of adventure stories as a point of departure to get some-where new, to invent new stories and construct new geographies and identities, to write new literatures. Critical adventures clearly depart from the stories and the geographies they refer to, but in many important ways they stay the same. The adventurer enters a *terra incognita* where anything is possible, where he or she is able to reinvent him or herself and his or her world. *Terra incognita* is an unsettling, disorienting space where it is not only possible but urgently necessary to remap, to invent new geographies. Unmapping, in the words of Robert Kroetsch (1989: 17), means 'unlearning so that we might learn'. Critical retellings of *Robinson Crusoe* by post-colonial writers such as Tournier re-enact *Robinson Crusoe* to the extent that they lead the Crusoe figure out of the world he knows, through a period of storm-like disorientation, to an unfamiliar world where nothing is certain but almost anything is possible. The less they carry with them into that unfamiliar space, the better chance they have of leaving the world as they know it behind. One thing they do need to hold on to is the ability to tell a good story, whether in a 'literary' or in a popular style. So long as they tell a good story, they keep the reader's attention while mapping a new world, or just defining space in which others may map their worlds. Reinscribing *terrae incognitae*, unsettling spaces in which to reinvent geographies and identities, post-colonial adventure writers continue rather than break with the radical tradition of their medium, in which worlds and world views are turned upside down before being recast in some other form. This tradition is as evident in the colonial adventure stories of Defoe and Marchant as it is in the post-colonial adventure stories of writers such as Selvon and Tournier.

8

CONCLUSION
Further adventures?

If the geography of adventure was ever a 'region of fun' in which readers enjoyed 'unbounded amusement', as Ballantyne (1858a: iii) hoped they would, it is now a problematic space, in which amusement is bounded and qualified. Adventure has become tainted with violent and repressive histories of imperialism, patriarchy and racism, partly through the efforts of intellectuals such as Edward Said and Ronald Hyam, whose critical and revisionist histories of imperialism have reached relatively popular audiences.[171] Contemporary unease with adventure is illustrated in an episode of 'The Modern Parents', published in *Viz* magazine in 1995. Like other features in *Viz*, which is currently the favourite magazine among British boys aged between 11 and 14 (Tucker 1995), 'The Modern Parents' is a textually self-conscious comic strip, replete with references – mostly satirical – not only to other comics, but also to comments on those comics. 'The Modern Parents' adventure sketch, a satire on academic criticisms of adventure, begins with the parents reading (the imaginary) Professor Erica Sharp's condemnation of *Biggles* as 'patriarchal, imperialistic, militaristic and racist propaganda of the worst kind'. In a dawn raid on the 'bedroom and meditation area' of Tarquin, their beleaguered offspring, the modern parents fill a recycling container with books, including such adventure stories as *Biggles Flies East, Tarzan of the Apes* and *Tintin in America* (Plate 8.1).[172]

The *Viz* sketch caricatures a dilemma: do we throw adventures (from books to computer games) into the recycling basket, or do we give in to their seductive violence and exoticism? The *Viz* solution to this dilemma is to parody and undermine the Professor's criticism of children's books, and to give in to the male pleasures of violent adventure. Following the dawn raid, the story proceeds in characteristic *Viz* style – the height of bad taste. The suggestion is that Professor Sharp, whose name labels her a woman and an academic, is out of touch with boyish desire for adventure and violence, which grips not only Tarquin but also his father. The humourless critics condemn adventure without offering an alternative. In place of his recycled stories, Tarquin is invited to read a book called *The Politics of Gender and E-Numbers*, and is dragged along to Professor Sharp's Green-Feminist Computer Cooperative. There are, it seems, no real alternatives to violent adventure. While clearly a caricature of the 'dilemma' facing readers

161

Plate 8.1 Dawn raid on Tarquin's collection of adventure books and magazines. One box from an episode of 'The Modern Parents' in *Viz* magazine, April/May 1995.

and writers of adventure literature today, the *Viz* sketch presents an important challenge: to steer a course between the uncomprehension of the recycling basket and the reassertion of violent adventure. This means acknowledging the popular appeal of adventure, and its resilience in the face of condemnation; it means not writing adventure off.[173] It means exchanging labels such as 'violent', 'imperialist' and 'racist' for a more sophisticated understanding of the politics of adventure. It means moving within the geography of adventure, both actively and critically, whether as readers or writers, or perhaps as some appropriately twentieth-century equivalent. It means asking where, in the geography of adventure, is it possible to go? Before asking where adventure can take us, I should say a little about where adventure has gone.

ADVENTURE TODAY

Adventure has been transformed since the end of the Second World War. On the surface, it may appear that adventure has declined. As early as 1940, Orwell commented that boys were beginning to find boys' weeklies such as *BOP* 'old-fashioned' and 'slow', and were looking elsewhere for entertainment (Orwell 1940: 91). By now, the boys' adventure books and magazines that used to flood book stalls and newsagents have largely disappeared, and many of their authors have been forgotten. But adventure fiction has shifted from one medium to another, and adventure non-fiction has proven equally resilient. The decline of juvenile adventure literature is partly a product of the decline of juveniles as a proportion of the population (Walvin 1982), and hence as a cultural influence. Whereas late nineteenth-century adults commonly read children's stories, the reverse is now true, as children read books and magazines intended primarily for adults. *Viz*, for example, is even labelled 'Not for sale to children'.[174] The decline of juvenile adventure literature is also a product of the decline in reading, and the corresponding increase in looking at pictures, watching television and film, and playing computer games. Twentieth-century technologies, in commercial context, have partially replaced printed words with visual images – although, as Joseph Bristow demonstrates in his 1995 anthology, *The Oxford Book of Adventure*, the printed adventure is far from dead.[175] Adventure, with its emphasis on setting and action, has been particularly influenced by changing narrative technologies, with their visual capabilities and drama. Journeys into the unknown (as well as into relatively familiar territory) are depicted, with primary emphasis upon photographic imagery, in the *National Geographic Magazine*, which is sold to more than eleven million people every month (Lewis 1985: 469), and seen by as many as fifty million (Abramson 1987: 5), both in the USA and around the world (see also Montgomery 1993; Rothenberg 1994). For the most part, the strong demand for geographical adventure info-tainment is met by commercial film-makers and other media.[176] While explorer and other adventure themes were popular in Victorian and Edwardian theatre (Bratton 1991), particularly in Britain, adventure drama did not reach mass audiences until the advent and popularisation of cinema, television and then video in the twentieth century (Shohat 1991). With the filming (and re-filming) of classic adventures, from *Tarzan* to *Around the World in Eighty Days* and *The Jungle Book*, and with the creation of new adventure stories based on old formulae, it would be difficult to argue that adventure has declined. Thus Brydon and Tiffin (1993: 41) argue that 'imperial fictions, old and new, still constantly inform popular culture'.

In this age of information technology and critical academic geography, one might imagine exploration and travel adventures by geographers and other serious non-fiction writers to be in decline. One might expect, fifty years after the President of the AAG observed that 'geographers seldom or never have the opportunity to enter literal *terrae incognitae*' (Wright 1947: 2) and directed them, instead, to the *terra incognita* inside their minds, that the adventure tradition in geography would be over. One might, perhaps, expect geographers to shy away from exploration and travel adventures, for fear of being associated with a men-only, dining-club, imperial sort of geography, which dominated the RGS in the late-Victorian and Edwardian period (Livingstone 1992). Yet the RGS director insists on calling the present time, not Victorian or Edwardian times, 'the golden age of discovery' (Lean 1995: 14). This in the year when the RGS effectively re-entered the mainstream of British geography, through a merge with the Institute of British Geographers (IBG). As the London *Independent on Sunday* reported in April 1995:

> Once again the panelled halls of the Royal Geographical Society, spiritual home of Stanley and Livingstone, Shackleton and Scott, are resounding to tales of discovery and derring-do.
>
> In the last few weeks the members of the society, which for the last 165 years has been the repository of the records of world exploration, have been notified of the discovery of the sources of two of the world's greatest rivers.
>
> (Lean 1995: 14)

The RGS still associates itself with travel and adventure, further, through corporate sponsorships with travel-related companies including Land Rover, British Airways and River Island.[177] Realistic exploration adventure stories, in modern form, are also set in space, in which astronauts are the new heroes (Bann 1990). In addition to the academic geographers who profess to narrate their own adventures, there are many more, such as Sauer who I mentioned in Chapter 1, whose work can be read as adventure literature. Livingstone (1992: 349) identifies a continuing tradition of exploration in the practice and vocabulary of academic geography. He observes that

> this expeditionary motif has become so engrained in the subject's collective memory that geographers have continued to speak of it in other contexts: expeditions into the urban jungle, ethnic ghettoes, and other threatening environments.

Also, 'serious' non-fiction writers and documentary film-makers have written and produced new adventure stories, many of them based on travel. Torgovnick (1990) identifies continuity in the ethnographic adventure tradition. In recent reincarnations, western encounters with 'primitive' peoples include, for example, Tobias Schneebaum's books about his encounters with Aborigines, and Lorne and Lawrence Blair's *Adventure* on American public television. The National Geographic Society, also, makes popular television documentary variations on the theme of voyages into the unknown. Non-fiction adventure, like its fictional counterpart, is very much alive, and the geography of adventure is still in cultural circulation.

AMBIVALENT SPACE

Adventures, like other maps, create conceptual space in which to move, without completely determining where the reader or writer goes. As points of departure, adventures are ambivalent.

Adventures set in the new world constructed imaginative geographies, points of departure. The European 'discovery' of a new world resulted in the transformation of both the new world and Europe itself, and popular adventure stories were, and in some cases still are, one of the mechanisms of that transformation.

Although Europe was not a closed system before Europeans encountered (what they later called) America, it was a very much more open – and hence dynamic – system after that moment. They had been 'on the move' for centuries before Columbus' famous voyage, but the so-called discovery of America transformed European prospects for exploration, trade and emigration, and initiated a series of encounters that were to have tangible consequences around the world. It created new possibilities for European geographical fantasies and stories, including adventure stories, which could be set in an expanded range of 'real' but 'unknown' settings. In the writing and reading of adventure stories, as in more tangible encounters and imperial practices, Europeans reconstituted themselves while, at the same time, they reconstituted the new world.

Their discoveries gave Europeans the chance to go somewhere else, in so doing transforming and reinventing both themselves and the spaces around them. In their dreams or in reality, they travelled to *terrae incognitae* where they sought to simplify and purify their culture. Christians, from the Puritans in New England to the Dukhobors in western Canada, left Europe for settings where they could purify, perfect and revitalise their religious culture. Like Robinson Crusoe on his island, they isolated and purified particular strands of European Christianity, while transforming a new world setting by colonising it. The story of Robinson Crusoe's transformation was addressed not only to potential colonists, but also to Christians who were to remain in Europe. Defoe's Christian adventurer, his faith purified away from European religious institutions, was an image of dissent, an argument for the deinstitutionalisation of religion. Other adventure stories, addressed primarily to European audiences, follow the same general pattern of isolating and nurturing particular elements of Europe in the spaces of adventure. In *The Young Fur Traders*, for example, Ballantyne imagined the simplification of European masculinity by isolating and naturalising one form of masculinity – Christian manliness – in a Canadian setting. Adventures can be located among a broad range of stories and histories in which Europeans, encountering the new world, profoundly transformed both Europe and the new world.[178]

The transformative capacity of the adventure story is rooted in its ambivalent mixture of conservatism and radicalism, its ability to map and remap – naturalising and fixing geographies and identities in realistic space – but also to unmap – subverting and destabilising received constructions of geography and identity. Adventure, although popular literature and therefore highly conventional, presenting what Northrop Frye (1990: 116) calls 'an unobstructed view of archetypes', is never a static or totally confining narrative. The geography of adventure, despite its 'appearance of reality' and hence of fixity, is never static. It is a constantly changing, magical world where reality is fleeting and almost anything is possible. The world of adventure presents what Robert Louis Stevenson called a 'kaleidoscopic dance of images' (Salmon 1888: 105). This is liminal terrain in which elements of the recognisable world reappear in strange, ever-changing, sometimes disturbing configurations. It is a space in which to move, not to stop. Like all geography – all maps – it both constrains and enables.

It constrains in the sense that it reproduces many values and assumptions, circumscribes some of the possibilities of geography and identity, and defines the terrain on which world views can be negotiated. It also enables in that it creates space in which writers and readers are able to rework and redefine values and assumptions, and begin to transgress boundaries and categories. The geography of adventure is a point of departure. As Patrick White put it in his metaphysical Robinsonade *Voss* (1957), 'Knowledge was never a matter of geography. Quite the reverse, it overflows all maps that exist' (White 1957: 475).[179] The geography of Voss's adventure is not the ultimate goal of his quest, but it is a necessary starting point.

Writers actively move within the spaces of adventure, reinventing and reinscribing social constructs such as race, class, gender and geography. The geography of *Robinson Crusoe*, for example, has been the point of departure for many mappings, unmappings and remappings. Just as some nineteenth-century writers (and editors) re-entered Defoe's eighteenth-century narrative to map nineteenth-century constructions of British colonialism in their many retellings and editions of *Robinson Crusoe*, late twentieth-century writers have re-entered the same narrative space to unmap those colonial constructions. They unmap, but not to bury the story. In most cases they continue to move within the general parameters of its geography. Since the geography of the island starts out as *terra incognita* and is transformed as the story progresses, as Crusoe transforms himself, the geography of the story is intrinsically dynamic and leaves scope for imagining alternative worlds. Within the general geography of *Robinson Crusoe* it was possible for colonists and emigration activists like Catharine Parr Traill and Bessie Marchant to map colonial geographies, and for post-colonial critics such as Sam Selvon and J.M. Coetzee to recover space in which people could invent post-colonial geographies and identities.[180] Similarly, academic and other 'serious' writers move within the geography of adventure, critically and self-consciously negotiating constructions of geography and identity. Sauer, for example, retold *The Early Spanish Main* as an adventure story, but a critical adventure story. The Columbian encounter was, in Sauer's narrative, an invasion by the modern, Western world, an assault on native ways. The invasion, involving 'exploitation' of natives and 'pacification by terror', decimated the peoples of the Spanish Main, leaving the region a 'sorry shell' of its former self (Sauer 1966: vii, 291, 294). Sauer is graphic and unequivocal. A relatively short time after the arrival of the Spanish, he writes,

> A well-structured and adjusted native society had become a formless proletariat in alien servitude, its customary habits and enjoyments lost. The will to live and to reproduce was thus weakened. One way out was to commit suicide by the juice of the bitter yuca.
>
> (Sauer 1966: 204)

Sauer's ability to tell a good adventure story, and his pride in the geographic tradition of exploration and adventure, did not prevent him from lamenting 'the first tragic confrontation of Europeans and American natives' that took place in the early Spanish Main. On the contrary, the adventure narrative, and the geography of adventure, provided Sauer with a medium in which to criticise European imperialism, and to rethink critically the Columbian encounter.

Readers and listeners, like writers and publishers, move within the narrative spaces of adventure. Readers select much of what they read, despite the censures of publishers, parents and teachers. Nineteenth-century children bought, swapped and read 'penny dreadfuls' against their parents' wishes, while their twentieth-century counterparts sought out and read stories by Enid Blyton, often against the wishes of their parents, teachers and librarians. As children's librarian and juvenile literature critic Sheila Ray (1982: 4–5) points out,

> Ever since children started reading, adults have been anxious to ban some of their favourite books. . . . *Robinson Crusoe* was criticised because the reading of it might lead to 'an early taste for a rambling life' and because it emphasised the importance of accumulating material goods. . . . No author, however, has been attacked to the same extent that Enid Blyton has during the last thirty years.

In particular, adults in post-war Britain objected to the racism, sexism and classism of Blyton stories, which seemed embarrassingly dated. But children 'have voted with their feet, making [Blyton] the popular author that she is' (Mullan 1987: 13). In addition to choosing what they read, readers choose how they read, sometimes receiving the images and messages intended by the author and publisher, sometimes actively challenging and reinterpreting them. One boy challenged the editor of *Nuggets* (a British magazine) to a duel! (Turner 1957). Most readers challenge authors and their texts less dramatically, inventing their own ways of reading them. With the potential to be read in different ways, adventure stories construct imaginative space but do not dictate what goes on within its broad parameters. As a result they afford readers a measure of agency, an active role in processes of constructing and reconstructing geographies and identities.

Politically ambivalent, the geography of adventure is both violent and liberating. Ideas like purification and simplification, common in 'new world' histories and stories – including adventure stories – can be violent when simplistic heroes and settings are mapped onto real human geography. Geographical descriptions, including narratives and maps, simplify and order the world. By definition, narratives are simpler than the world and the events they describe. A typical dictionary definition of narratology (Gray 1992: 191) explains that:

> A basic idea of narratology is that a 'story', the succession of events that are to be recounted, in their simplest chronological form, is manipulated in a variety of different, definable ways into a finished narrative, which is the text.

As Michael Mann (1986: 4) put it, 'Societies are much *messier* than our theories of them.' But the violence of narrative and map, if indeed there is any, is not intrinsic to narratology and cartography. Narratives and maps become violent when literalised, mapped directly onto real people and places. When a simple ideology of manliness, articulated in the uncomplicated setting of a boys' adventure story like *The Young Fur Traders*, is literalised as a realistic and plausible representation of manliness, the heterogeneity and difference among boys and men is denied; all are forced, violently, into a narrow, uncomfortable box. As Graham Dawson shows in *Soldier Heroes*, a Kleinian interpretation of relationships between adventure heroes and readers, identification with heroes involves violent 'psychic splitting' in which readers' psychological wholeness is compromised in favour of coherent, recognisable masculine identities. Violence is done to human geography, more generally, when ideologies of purified religion and simplified culture, articulated in stories like *Robinson Crusoe*, are mapped onto real space. *Terra incognita* is not discovered, it is socially produced through the erasure of human geography (Shohat 1991). Before the new world could become the *terra incognita* for Europeans to purify their religion and simplify their culture, its human geography – people, flora and fauna, stories and maps – had to be erased, often violently. And before Europeans could imagine sweepingly simple, brave new worlds in their own continent, it was necessary for them to erase the worlds that existed – a violent prospect, for as Foigny's adventure story made clear, brave new worlds rise out of ashes. Foigny's utopia rose from the ashes of the *ancien régime*, while twentieth-century utopias and dystopias emerge from the ashes of the modern world.[181] But while ideas like simplification and purification are violent when mapped onto real people and places, they are still potentially liberating. The geographies of adventure, simple and uncomplicated, enable writers and readers to remove themselves from the messy realities and textured experiences of here and now, enabling them to imagine alternatives, other possible worlds, departures from the *status quo*. There are limits, however; adventure does not set the imagination completely free.

THE LIMITS TO ADVENTURE

While adventure frees the imagination, opening up material and metaphorical space in which to move, the geography of adventure is ultimately bounded, and the imagination is ultimately confined to a limited range of politics. Like the 'cognitive maps' proposed (but not defined) by Jameson, geographies of adventure are relatively open-ended, not deterministic. But, like Jameson's cognitive maps, which will accommodate a new socialist politics, the spaces of adventure are geared towards certain politics and away from others. In the post-colonial context, adventure stories are mostly either allied with imperialism or critical of imperialism, in a reactive way. Since imperialism did not disappear with the fall of British and other European empires, but re-emerged in various late twentieth-century guises, resistance to imperialism is as important and as urgent as it ever was (Said 1993; Smith 1994). But to resist imperialism, in the same conceptual space where imperialism was constructed, is to resist reactively rather than actively. Thus Selvon and Coetzee, recognising the limitations of inherited and appropriated conceptual space, attempt to deconstruct that ground, to move into new kinds of space where new kinds of politics are possible. And thus Jameson, envisioning a new politics, seeks not to appropriate, but to create conceptual space. However, a difference between spaces of adventure, which may function as post-colonial sites of resistance, and spaces of Jameson's new Socialist politics is that the former exist. The history of adventure is a history of gradual change, in accessible imaginative space that grips the popular geographical imagination. Its boundaries, while very real, are constantly shifting. Slowly changing, the geography of adventure is always associated with conservative constructions of geography and identity, yet always associated with the possibility of change, the dream of something and somewhere else.

NOTES

1 INTRODUCTION

1 I borrow the phrase 'into the unknown' from British Victorian adventure writer Rider Haggard, who used it as the caption to an illustration in one of his African adventure stories (Springhall 1973: 1525).

2 Dates of first edition are indicated in brackets the first time each book is mentioned. First editions of primary sources are normally cited thereafter.

3 Anonymous introduction to *Treasure Island* (Stevenson 1962: 3). Macgregor (1989: 19) writes that the map was drawn by, rather than for, Stevenson's stepson Lloyd. *Treasure Island* was initially published in serialised form as *The Sea Cook*, by Captain George North (Turner 1957). Stevenson (1850–1894) lived and wrote in Edinburgh, but travelled widely.

4 In colonial cinema, similarly, adventures commonly begin with images of maps, in which *terra incognita* is prominent (Shohat 1991).

5 On Victorian ideals of childhood innocence, see Bratton (1990) and Nelson (1991).

6 In the introduction to the twelve-volume atlas, *Le Grand Atlas* (1663) (Alpers 1983: 159). The atlas is discussed at greater length by Koeman (1970).

7 I use the term 'modern' to refer to an historical period, beginning in this case with the first European 'discoveries' in (what became) America at the end of the fifteenth century, and continuing through the present. I refer to a period, not to any intrinsically 'modern' cultural, economic or political form.

8 I refer to Britain rather than England, since the Empire was British rather than English, and since Scots were prominent in the writing, publishing and reading of adventure stories, as they were in practical acts of colonisation. I refer mainly to the period when Scotland had become united with England and Wales, in Great Britain.

9 Hermann Ullrich coined the term Robinsonade in 1898, in the title of his bibliographical study *Robinson und Robinsonaden*. French critics, following Ullrich, use the term 'la robinsonnade', a form of *voyage imaginaire* (Green 1990).

10 In Britain, *Moby Dick* was originally entitled *The Whale*.

11 Like Ransome and most other readers, I normally refer to adventure stories by their short titles. Full titles are listed in the bibliography and, where relevant, in the text.

12 As Keltie (1907: 191) described *Robinson Crusoe*.

13 As Leighly (1963: 4–5) observes, 'The words "pioneer" and "frontier" are among the ones that occur most frequently in Sauer's writings. In his pages people are usually on the move or establishing themselves in new surroundings. They burn the grass and trees, plant in newly cleared woodland, break the prairie sod, seek the best places for their houses.'

14 It was common for Victorian and Edwardian adventure heroes to declare their distaste for book learning and school in general. Conrad's views on school geography can therefore be interpreted, at least in part, as part of his credentials as an adventure hero, rather than a literal or specific comment on 'real' Polish or British geography lessons.

15 The examples of New Zealand and the United States of America are selected on the basis of availability of survey data.

16 For example, Major (1991) remarks on Marchant's disappearance from public view in Britain, while Harrison (1980) traces the general decline of adventure and other popular fiction in western Canada.

17 Vincent, using signatures (and marks) on the marriage register as an indicator of literacy, concluded that 1839 illiteracy rates of 33 per cent for grooms and 49 per cent for brides were wiped out by 1914. In the registration districts he sampled, illiteracy was reduced to 1 per cent by 1914. The working classes made the greatest gains in this period since it was they who, initially, were less likely to be literate. (Vincent 1989; Stone 1969).

18 Turner's liminal space is metaphorically related to the spatially and temporally detached 'zone unknown', described by anthropologist Arnold Van Gennep (1909, 1960), in which tribal rites of passage (such as the passage from childhood to adulthood) take place.

19 As the metaphorical relationships with literally liminal processes (tribal rites of passage) become stretched, the term liminal is less accurate, and liminoid is preferable (Turner 1982). Liminoid spaces are often associated with leisure and leisure time, themselves products of industrialisation, with its spatial and temporal divisions between work and play, and its mass-consumption of leisure activities and products.

20 This quotation is from the second page of Defoe's unpaginated preface. Defoe elaborated this qualification (of his realism) further in *Serious Reflections* (1720), the second sequel to *Robinson Crusoe*.

21 I qualify this reading of *Robinson Crusoe* and other adventure stories below, suggesting the possibility of alternative readings and retellings.

22 For purposes of clarity, I use the terms Canada and Australia throughout, although these countries have, of course, previously been known by different names, and have been constituted differently, geographically and politically.

23 I put caution marks around the terms masculinist and imperialist, terms which imply universal constructions of masculinity and imperialism, while masculinities and imperialisms are historically constructed, not universal.

2 MAPPING ADVENTURES

24 Atkinson uses the terms imaginary and extraordinary interchangeably. The more technical term is 'philosophic adventure novel in a realistic setting' (Gove 1975).

25 Although they were not known as Robinsonades, since the term was not coined until 1898. The plot of *The Isle of Pines* is summarised in its subtitle, *a Late Discovery of a Fourth Island Near Terra Australis Incognita, by Henry Cornelius van Sloetten, Wherein is Contained a True Relation of Certain English Persons, who in Queen Elizabeth's Time, Making a Voyage to the East Indies Were Cast Away, and Wrecked Near the Coast of Terra Australis Incognita, and all Drowned, Except One Man and Four Women, and how Lately Anno Dom. 1667 a Dutch Ship . . . by Chance have Found their Posterity, (Speaking Good English,) to Amount, (as They Suppose,) to Ten or Twelve Thousand Persons.* This story 'is often viewed as the first Robinsonade prior to Defoe's work' (Rigney 1991: ii). The existence of

Robinsonades prior to 1719 underlines the point that Defoe's novel was a landmark but not an entirely original work, and that it had antecedents. The term Robinsonade is a retrospective, critical term, which authors like Neville and Defoe did not use.

26 This story of Alexander Selkirk's shipwreck was written by Captain Woodes Rogers, commander of *The Duke*. Other versions of the story were penned by Edward Cook, second captain of *The Duchess*, and by an anonymous pamphleteer, and by Sir Richard Steele (Baker 1931).

   More generically, Defoe's shipwreck in the new world borrows from a wider European tradition that reaches back at least as far as Shakespeare's 'American' play, *The Tempest*, which Leo Marx (1964: 72) reads as 'a prologue to American literature'. Shakespeare's storm is thought to have been inspired by the account of a contemporary traveller, William Strachey (Marx 1964; Gillies 1994).

27 For example, Defoe also took liberties with received anthropological 'facts'. His cannibals were based on the Caribs, after whom the Caribbean Sea is named, although Caribs did not engage in cannibalism merely for sustenance, and did not eat just anyone they could kill. Their cannibalism was ceremonial, religious and selective (Pearlman 1976).

28 Britain did not begin to approach universal literacy until the turn of the twentieth century, although mass literacy was achieved earlier in the Victorian period (Vincent 1989).

29 Friday and Xury.

30 Critical attention to *Robinson Crusoe* is extremely voluminous and defies easy summary, although its range is illustrated in anthologies and readings by Shinagel (1975), Rogers (1979) and Seidel (1991).

31 Then, as now, geography was a 'contested enterprise' (as Livingstone 1992 puts it) and the geographical fraternity was correspondingly fragmented, encompassing investors, explorers, and various scholars, gentlemen and other writers and mapmakers.

32 Not all geographers, of course.

33 Earlier European colonial settlements, antecedents of the Victorian mass-settlement schemes, are discussed by Hartz (1964) and Harris (1977).

34 Nerlich (1987: 2. 263) argues that adventure was variously a 'courtly-knightly' ideology (twelfth-century France) and a mercantile ideology (fifteenth-century England), before Defoe wrote *Robinson Crusoe* and made adventure 'the ideology of the middle class'.

35 On the abridgement of *Robinson Crusoe*, see Dahl (1977). The redefinition of *Robinson Crusoe* as juvenile literature was partly thanks to Rousseau (1762), who recommended the book to children and said it was the most important, if necessary the only book they should read. A century later, Leslie Stephen (1874: 56) argued that due to 'the want of power of describing emotion as compared with the amazing power in describing facts, *Robinson Crusoe* is a book for boys rather than men, and, as Charles Lamb says, for the kitchen rather than for higher circles'.

36 Tessa Chester, of the Bethnal Green Museum of Childhood, dates the book between 1830 and 1836 (personal communication, January 1996).

37 Chapbook editions, commonly around thirty or forty pages, shrank to as few as eight. One such edition, sold for a farthing, is comprised of a folded sheet, divided into eight pages, each 17 by 11 centimetres. It is illustrated haphazardly with various illustrations tacked together. It is undated and anonymous, published by R. March, London. It is part of a collection of children's books held at the Bethnal Green Museum of Childhood, London.

38 On the illustration of *Robinson Crusoe*, see Wackermann (1976).

39 But there are also differences between *Robinson Crusoe* and the tribesmen and women described by anthropologist Arnold van Gennep, who inhabited the most literally liminal space (Turner 1969). In particular, the former is a fictional character while the latter are real people.

40 The story of Crusoe's spiritual rebirth, emerging from the dark waters into an unnamed land, a garden of Eden where he is the sole, Adam-like occupant, is reminiscent of the Biblical creation story. In Genesis, land creatures including Adam and Eve were created on the sixth day, once the natural environment was in place (night and day; sky; land and vegetation; sun, moon and stars; sea creatures and birds). In other words, Adam appeared on Saturday while (most of) nature appeared by or on Friday.

41 I borrow this phrase from Paul Carter (1987: 17), who traces the emergence of 'the shadowy outline of a place' in Australian exploration literature in *The Road to Botany Bay*.

42 Crusoe takes his capitalist enterprise as far as circumstances permit. Due to his isolation, he is unable to participate in a market economy, and is therefore forced into a more self-sufficient, superficially Jeffersonian form of colonialism.

43 Defoe criticised the brutality of some slave traders, and condemned the more plunderous forms of imperialism, although he was not opposed to slavery or imperialism *per se* (Rogers 1979; Seidel 1991). The 'wickedness' of slavery, identified in *Robinson Crusoe*, was associated less with the suffering of slaves than with the moral effects on white colonists, who might gamble capital and live fast rather than sober lives.

44 Schieder (1986) provides the most comprehensive bibliographical history of *Canadian Crusoes*. The translation of *Welsh Family Crusoes* had a wider circulation than its predecessor.

45 Slightly earlier Robinsonades, such as Johann Wyss's *Swiss Family Robinson* (1814) and Captain Marryat's *Masterman Ready* (1841), included mixed casts of adults and children.

46 Ballantyne's debt to Copley was drawn to my attention by Vaughan Cummins, of the School of Art, University of Wales, Aberystwyth.

47 Mango, one of the islands on which Ballantyne's story is set, is also the name Columbus gave to Cuba, which he mistook for Mangi, South China (Sauer 1966).

48 Ballantyne endorses the Christian mission, but not this particular trader, who might better be labelled a pirate.

49 The pattern of Canadian settlement has been compared, more generally, to an island archipelago by historical geographer Cole Harris (1982).

50 The archives of nineteenth-century Canada are replete with reports and stories of lost settlers. For example, the *Regina Leader* (4 August 1898, p.5 column 4) reported, 'Lost on the Prairie – Mrs David Mackenzie of Little Hay Lakes near Wetaskiwin, Alta., an old lady nearly 70 years of age, got lost while hunting for some cattle. . . . She wandered around for 5 days without a thing to eat.' The lost child, a recurring motif in Canadian colonial literature, also caught the imagination of Catharine's more famous sister, who wrote of 'lost children' in *Life in the Clearings Versus the Bush* (Moodie 1853).

3 MAPPING MEN

51 I refer to Sauer's reading of geographical literature (Sauer 1925: 21, my emphasis), which I have called adventure literature.

52 This means that critical geographies of masculinist adventure are allied, both

NOTES

intellectually and politically, with feminist, anti-racist and other attempts to de-naturalise hegemonic masculinities, which 'concur in affirming the social and cultural construction of masculinity' (Roper and Tosh 1991: 8; also see Boyd 1991; Brantlinger 1988; Bratton 1989; Dawson 1994; Mackenzie 1986; Reynolds 1990; Richards 1989).

53 After three editions Ballantyne's original title was shortened to *The Young Fur Traders*. For the sake of clarity, I will use this title to refer to all editions.

54 One such fan, a 15-year-old boy who rushed up to the author in a public park, expressing admiration for his stories and inviting him to dinner, was Robert Louis Stevenson (Quayle 1967).

55 Comparable data not available for the 1850s, when *The Young Fur Traders* was first published.

56 The British Library catalogue lists twelve editions of *The Young Fur Traders* between 1856 and 1950. These were published by Nelson (1856, 1857, 1901, 1908); Ward, Lock & Co. (1901); Nisbet (1901); Partridge (1913); Oxford University Press (1923); Blackie (1925, reissued in 1950); Juvenile Productions (1937); and Gawthorn (1948). The book has since been printed occasionally. In 1979, for example, it was included in *The Coral Island; and, The Young Fur Traders*, published by Octopus (London).

57 Ballantyne's other North American adventures include *Ungava* (1858b); *The Dog Crusoe* (1861a); *The Golden Dream* (1861b); 'Away in the Wilderness' (1863b); *The Wild Man of the West* (1863a); 'Over the Rocky Mountains' (1869); *Tales of Adventure by Flood, Field and Mountain* (1874); *The Red Man's Revenge* (1880); *The Big Otter* (1887) and *The Buffalo Runners* (1891). Quayle (1968) gives a comprehensive bibliography of the first editions of R.M. Ballantyne.

58 Egoff (1992) provides an extensive bibliography of children's books set in Canada published before 1940.

59 John Newbery, whose *The History of Little Goody Two-Shoes* (1765) ran to approximately 200 editions in Britain and America, is sometimes credited with inventing children's books. He was, at least, the first to promote juvenile literature as an important branch of literature (Barr 1986).

60 Of the 790 boys surveyed, twenty-four said Defoe was their favourite author or writer of fiction; forty-three said *Robinson Crusoe* was their favourite book. Of the 1,210 girls, eight said Defoe was their favourite author or writer of fiction, while two said *Robinson Crusoe* was their favourite book.

61 In an Australian oral history of reading in the period 1890 through 1930 (Lyons and Taksa 1992).

62 In its 25th annual report (1824), the RTS acknowledged that its 1,688,760 children's books sold the previous year represented no more than one-fifth of the market (Bratton 1981: 38).

63 *By Track and Trail*, written and illustrated by Roper, F.R.G.S., was one of the many transcontinental travel and adventure narratives produced in the decade following the completion of the transcontinental railroad in 1885.

64 Adolescence is said to have been 'discovered' in Britain and America in the Victorian period, particularly the 1890s (Demos and Demos 1969; Kett 1971, 1977), although historians do sometimes speak of adolescence in earlier contexts (Ben-Amos 1995, for example). While the concept of adolescence was to be applied universally – to an entire age group – early in the twentieth century (Hall 1904), its origins in the nineteenth century were more class and gender specific. Kett (1977) argues that adolescence, associated with increased schooling and deferral of employment and other responsibilities of adulthood, was a middle-class phenomenon in North America and Britain in the 1880s and 1890s. Dyhouse

(1981) argues, further, that the concept and phenomenon of adolescence was predominantly male, since it was associated with male privilege, particularly schooling and the gradual acquisition of independence. She argues that female adolescence, like male, began as a middle-class phenomenon.

65 Arrowsmith was effectively the official, definitive cartographer of western Canada in the mid-nineteenth century. Its maps of the region, regularly updated between 1834 and 1858, are dedicated to the HBC, and contain 'the latest information which [HBC] documents furnish'.

66 Paul Carter's discussion of Australian explorers' place names in *The Road to Botany Bay* (1987) parallels mine of adventurers' place names. Explorers, he argues, preserve the traces of encounter in their maps and names, which serve as directional pointers in dynamic landscapes, rather than geographic 'facts' in more static landscapes.

67 Cooper read about the western prairie in Lewis and Clark's journal; he never travelled there himself (Milham 1964).

68 In other stories, Ballantyne (1861a: 79) claims that 'it is scarcely possible to conceive a wilder or more ferocious and terrible monster than a buffalo bull', and describes a grizzly bear as 'one of the most desperate monsters and most dreaded animals on the face of the earth' (Ballantyne 1863a: 50). The fight with a wounded bear, relatively short and bloodless in *The Young Fur Traders*, is longer and bloodier in many later Victorian stories (including Ballantyne's), in which gratuitous violence had become more familiar and more acceptable.

69 Red River was the agricultural settlement, populated mostly by whites and Metis, located a short distance from HBC Fort Garry, where the city of Winnipeg now stands.

70 Down-market magazines were guilty of many geographical 'errors' (Turner 1957: 107).

71 Other interpretations of girls' and women's responses to male adventures are examined below, in Chapters 5 and 6.

4 MAPPING EMPIRE

72 Dixon (1995) provides a fuller account of Australian exploration and other adventure stories.

73 By Australian content, I refer to images and ideas specific to Australian people and places, and to input from people (such as travellers and settlers) and organisations (such as publishers and publications) in Australia.

74 Kingston retold exploration as non-fiction juvenile adventure in books such as *Captain Cook, His Life, Voyages, and Discoveries* (1871), while G.A. Henty (Arnold 1980) and others fictionalised exploration adventures, in books like *The Pioneers* (Ballantyne 1872) and *North Overland with Franklin* (Oxley 1901).

75 Since European settlement began, a 'high proportion' of Europeans have lived in urban centres (Camm and McQuilton 1988: 102). By 1861, over two-fifths of the European population lived in urban areas (centres of one thousand or more), and this proportion rose to over half in 1881 and over three-quarters in 1947.

76 Leichhardt, who had successfully explored parts of north-eastern Australia on two expeditions in the mid-1840s, set out in 1848 to cross the continent from east to west, but disappeared without a trace. Robert O'Hara Burke and William John Wills attempted to cross from Melbourne to north Australia in 1860, but died on the return journey the following year, along with many of their party.

77 What is 'useful' is, of course, culturally and historically specific.

78 Pastoralists encroached into most of the south-eastern area known as the 'fertile crescent' by the middle of the nineteenth century, then colonised the much larger areas of interior Queensland, New South Wales and South Australia before the end of the century (Camm and McQuilton 1988).

79 In 1841 Edward John Eyre made the first successful east–west crossing, along the coast of the Great Australian Bight. In 1844 Charles Sturt explored the area of desert and salt lakes north of Adelaide. Early nineteenth-century explorers John Oxley and Thomas Mitchell set out from Bathurst, New South Wales, and explored the tributaries of the Darling.

80 Writers of exploration narratives, as of all adventure narratives, follow literary models, whether they do so self-consciously or not (Davey 1979; Hodgson 1967; Knox-Shaw 1987; MacLulich 1979; Riffenburgh 1993). The literary models that shape *The History* were mostly supplied by early nineteenth-century Australian exploration journals, such as those of Oxley and Mitchell, which in turn, were indebted to inland exploration narratives such as Nicholas Biddle's account of the Lewis and Clark expedition (in North America) (Dixon 1986).

81 'When Sturt,' for example, 'speaks of the awful temperature that rendered life unbearable, and the inland slopes of Australia unfitted for human habitation', he is not quite himself. Like most of his party, he is 'weak and suffering', oversensitive to 'oppressive heat or extreme cold' (Favenc 1888: 141).

82 Favenc's career as a geographical writer, while mixed, was not exceptional. Others who turned from formal geographies to adventure and mystery, and from 'serious' adult works to juvenile romances, include Alexander Macdonald. A contemporary of Favenc, Macdonald (1828–1917) was an overlander and surveyor, an academic geographer who published papers in geography and founded the Victorian Branch of the Royal Geographical Society of Australasia (Macdonald 1915), and also a writer of adventure fiction.

83 Mysterious cave drawings, attributed to some civilised race, had been seen by George Grey, and were reproduced in an appendix to *The History* (Favenc 1888).

84 Street identifies the late-Victorian and Edwardian period as a time of unprecedented racism. His study of images of race and racism in British popular culture, *The Savage in Literature* (1975), begins with the publication of *The Origin of Species* in 1858, continues through the decades in which Europeans 'scrambled' for Africa, and ends after the First World War when, he argues, Malinowski and others transformed English anthropology, substituting biological and racial conceptions of culture with more relativistic and engaged studies.

85 John Crawfurd, speaking in 1861 on 'The connexion between ethnology and physical geography', cited by Livingstone (1992: 222).

86 Imagining Australia, Favenc continued a literary tradition that dates back to seventeenth-century French and eighteenth-century English imaginary voyages (see Chapter 6) (Gibson 1984, 1992).

87 The importance of the 1890s as a moment of origin has perhaps been over-stated, since bush mythology has earlier and more disparate origins (Lansbury 1970), and since the 1890s tradition is in part a creation of twentieth-century critics (Burn 1980; Schaffer 1988).

88 Founded in 1880, achieving a circulation of over 80,000, and publishing such writers and poets as Henry Lawson and Joseph Furphy, *The Bulletin* was the mouthpiece of nationalists, and has been called 'the chief instrument for expressing and defining the national being' (Palmer 1954: 92).

89 For example, the dust jacket of a children's history of Australian exploration explains that '*Doctor Hunger and Captain Thirst* may be written about Australian

history but it is really a series of adventure stories' (Hooper 1982). Hooper retells exploration as adventure.

90 When contemporary critics of Taylor – notably the outspoken Icelandic-Canadian, V. Steffansson – argued that Australia had great settlement potential, they still spoke of a generic national landscape.

5 AMBIVALENCE IN THE GEOGRAPHY OF ADVENTURE

91 Women's travel and travel writing are discussed, with reference to gender and imperialism, by Birkett (1991), Blunt (1994) and Blunt and Rose (1994). Domosh (1991) comments on relationships between women and exploration, Stanley (1995) on women and piracy. On constructions of femininity in girls' literature, see Bratton (1989), Thomson (1956) and Tinkler (1995).

92 Marchant also wrote several boys' adventure stories, such as *Athabasca Bill: a Tale of the Far West* (1906), which was published by the Christian Knowledge Society, London.

93 Blackie was Marchant's principal publisher, publishing 66 of her 150 novels (Major 1991).

94 Marchant's other Canadian stories include *A Countess from Canada* (1911), *Rachel Out West* (1923), *A Canadian Farm Mystery* (1917) and *Cynthia Wins* (1918), all of which were published by Blackie, London.

95 Although not all women and children could afford to conform to this effectively middle-class ideal; some went out to work, mostly as domestic or industrial labourers.

96 Francke, in a film review entitled 'Danger – Boy Zone' – a reference partly to the all-male world of *Boy's Own* – Lizzie Francke (1995: 26) calls *Crimson Tide* 'an excuse for men to get all hot under the collar together' and, she admits, 'this femme, at least, loves every inch of it'. For her, and 'for the women with whom [she] went to see it', who 'had all had a bad day at work', the 'slick fantasy of mean machine power perfectly fitted the mood'.

97 Adventure stories influenced many prospective emigrants. In the peak immigration years between 1897 and 1913, three-quarters of immigrants to western Canada were British or of British descent (Friesen 1984), and the majority of these were literate. They were exposed to, and presumably influenced by, the images of western Canada they encountered in popular literature, ranging from emigration propaganda to adventure stories (Francis 1989; Harrison 1980; Moyles and Owram 1988).

98 Adventurous emigrants, inspired by boys' stories, must have been disappointed and bored by the farm life that most of them found in Canada.

99 The bones were gathered together, removed and sold as fertiliser, so that soon the only reminders of their existence were the buffalo wallows, depressions in the prairie.

100 H.Y. Hind (Canada) and Capt. J. Palliser (Britain) led expeditions to western Canada in 1858. Palliser reported the existence of a 'fertile belt' stretching across the northwest, bounded by a frozen north and an arid south (Palliser 1859). Climatologist Lorin Blodgett, writing for a popular audience at about the same time, constructed isotherms that redefined parts of the northwest as decidedly temperate (Dunbar 1973). The optimism of Hind, Palliser and Blodgett was subsequently surpassed by that of 'experts' such as John Macoun (celebrity botanist) and George Mercer (popular geologist) who went on to question and qualify the idea that the southern prairie grassland was desert-like. Macoun claimed the prairie region was one big fertile belt, bounded by an arid southern limit that corresponded

roughly with the 49th parallel (the international border) and by a cold northern limit somewhere in the vicinity of the Arctic circle.

101 For example, Daunt (1885: 113) reminded readers that his hunting adventure *The Three Trappers* was set in the past, and predicted that 'in a few years the romance of the prairies will be a tradition of the past'.

102 British women's emigration societies, the first of which was founded in 1862, promoted and assisted women emigrants, and favoured Canadian destinations. The first such society, The Female Middle-Class Emigration Society, was founded by Jane Lewin and Maria Rye, after they were inspired by Susanna Moodie's 1854 settlement narrative, *Roughing it in the Bush* (Jackel 1982).

103 Ballantyne's imaginative geography was echoed in travel writing and fine art, by writers and artists such as the Earl of Southesk (in the late 1850s) and Paul Kane (in the 1840s). Other than in some vaguely nostalgic undertones, these artists did not acknowledge the (beginning of the) HBC's decline, or the changes that were beginning to take place in western Canada. Conventionally Romantic travellers, they sought out ancient and natural-looking elements in the landscape. Kane portrayed Indians and buffalo in *The Wanderings of an Artist* (1859) and in paintings, some commissioned by the HBC (Harper 1971). Southesk focused on craggy peaks and wild beasts, which inspired his poetic imagination.

104 Horner's parody of adventure begins with an apparent robbery, but ends in anti-climax when it is found that the 'thief' is nothing but a large rat. The protagonists – a widower and his 12-year-old son – are helpless in the absence of women, and they inhabit an all-male world that is more squalid than romantic.

105 Stead was a publicity agent in the CPR Department of Natural Resources (1912–1916) at Calgary; in 1919 he became director of publicity in the Ministry of Immigration and Colonisation in Ottawa.

106 Begg set his anti-adventure novel *Dot it Down* (1871) in a civilised, cultivated and prosperous Manitoba, and appended an 'Emigrant's Guide to Manitoba' in case any readers should miss the point.

107 Stead to T. Fisher Unwin, 14 Nov. 1912, Public Archives Canada MG30 D74 1–9.

108 James Macdonald Oxley (1855–1907), who was born in Nova Scotia and educated in law, wrote 31 books for boys (for example Oxley 1892, 1893), most of them set in Canada and published in Canada and the USA (Egoff 1992: catalogue entry #305).

109 Kingston 'toiled horribly' for empire, promoting emigration in the adventure stories and emigration pamphlets he wrote, and serving as secretary of the Colonisation Society, in which capacity he lobbied for increased attention to, and expenditure on, emigration (Kingsford 1947: 73). Kingston depicted emigration fields in adventure stories. His western Canadian adventures, for example, are set in a 'fertile belt' destined to be 'covered with populous towns and villages, and flourishing farms' (Kingston 1879: 2).

110 Jessie Saxby (1842–1940), a champion of women's emigration, visited the north-west in 1888 and 1889 (Peel 1968: 1–2). She promoted the region as a destination for female emigrants in *The Edinburgh Scotsman* and in adventure stories, notably *West-Nor'West* (1890), *Brown Jack* (1896), and *Kate and Jean; the History of Two Young and Independent Spinsters* (1889).

111 Argyll Saxby, a Scottish emigrant who settled in western Canada, wrote settlement adventures including *The Call of Honour* (1907), *Braves, Whites and Red* (1907), and *The Settler of Serpent Creek* (1921).

112 Oxley made a number of applications for financial assistance from the CPR, including one in 1890 to write up a CPR tour, although his requests were rejected

(Oxley letter to W.C. Van Horne, 12 September 1890, Van Horne Correspondences #30006, Canadian Pacific Corporate Archive, Montreal). Nevertheless, *The Boy Tramps* reads like CPR propaganda. The boys learn, for example, that 'there's no man deserves more credit for [the CPR construction] than him that's now president of the road' (Oxley 1896: 201).

113 Others who introduced sedentary, Christian Indians into their adventure stories include John McDougall, an Ontario Methodist minister, missionary and adventure writer who supported Christian mission and agricultural settlement on the prairies (McDougall 1895), and Egerton Ryerson Young, who attempted to depict 'genuine Indians', members of 'civilisation' rather than *dramatis personae* in 'scalping scenes' (Young 1899: 377). He claimed in his stories 'that boys could have such a jolly time with a lot of Christian Indians' (Young 1899: 5).

114 Fertile-belt rhetoric is echoed similarly in other adventure stories, ranging from Kingston's *Frontier Fort* (1879) to Milton and Cheadle's *North-West Passage By Land* (1865).

115 British Columbia joined Canadian Confederation on the condition that a transcontinental railroad would be built within ten years. The CPR, which linked western and eastern provinces, was completed and opened in 1885.

116 At 17 years old, Nell is no longer a child. In the Victorian period, at least, childhood was generally over by the age of 14 (Walvin 1982). If not a child, Nell is adolescent rather than adult, according to Edwardian definitions of adolescence. While adolescence was largely the privilege of middle-class boys in Britain and North America throughout most of the Victorian period, it was extended to include the entire teenage cohort, regardless of educational or employment status, in the first and second decade of the twentieth century (Dyhouse 1981). Since Nell is adolescent, passing from late girlhood to early womanhood during the course of the story, I refer to her as both girl and woman in the course of this chapter.

117 '"You can't go alone – you are only a girl!" he exclaimed, dropping a handful of spoons with a clatter because he was so amazed at the daring and audacity of Pam's great idea' (Marchant 1917: 13).

118 Bratley Junction and Camp's Gulch are fictitious places, although Lytton and New Westminster are real places, on the main line of the CPR and at its terminus, respectively.

119 This tradition of Amazonian heroines in adventure literature continues in the twentieth century, most famously, in the figure of Enid Blyton's *Famous Five* character George, whom her companions sometimes called a 'pretend boy' (Mullan 1987: 74).

120 R. Lindsay, letter to W.C. Van Horne, 28 Dec 1883, Van Horne Correspondences #2604 (Canadian Pacific Corporate Archive, Montreal).

121 In this literary landscape, which Kroetsch traces from popular adventures through to prairie realist fiction and other twentieth-century literatures of the Canadian north and west, horses exist in dialectical opposition to houses, and men in tense, dialectical opposition to women. Here, caricatured masculinities produce distance between women and men, charting space for men and boys but leaving no room for girls and women (Kroetsch 1989: 79–81).

6 READING AND RESISTANCE

122 Genet's child psychologist reported that his patient possessed a 'dubious mentality, abused by the reading of adventure novels for which, it seems, he was very avid' (E. White 1993: 48). In his biography of Genet, Edmund White (1993) suggests that

adventure books were among Genet's principal influences, in his life as in his plays and books.

123 The literary term 'imaginary voyage' was not used before 1741, although many imaginary voyages were published before that date, not only in France but also in other European countries, principally England and Germany. Of the 215 eighteenth-century imaginary voyages in Gove's annotated bibliography, 67 are English, 65 French, 59 German, 10 Dutch, 5 Danish, 5 Swedish, 2 Italian, 1 Latin, 1 Japanese (Gove 1975: 176).

124 The 1692 edition was probably the work of the French deist François Raguenet. The elements of the story that I pay greatest attention to in this chapter – the perilous journey and the adventures surrounding Sadeur's arrival in and departure from the broadly defined Australian utopia – are essentially the same in the 1676 and 1692/1693 editions. The precise details of the Australian utopia, which I am less concerned with, were modified by Raguenet. The 1676 edition has been translated by Fausett (Foigny 1993).

125 Garagnon (1981) interprets French imaginary voyages as anti-monarchist and revolutionary.

126 British counterparts punished for writing extraordinary voyages include Henry Neville, who was banished, and Bishop Hall, who was imprisoned in the Tower of London (Mackannes 1937; Wands 1981).

127 The word 'utopia' was coined by Thomas More in 1516 (Johnson 1981).

128 Sadeur initially refuses 'the call' to adventure, as adventure heroes traditionally do (Campbell 1949). For example, in order to avoid a sea voyage, as a child, he travels overland. But he is soon captured by pirates who have come ashore, and who take him to sea.

129 More explicitly, Savinien's imaginary voyage to the moon begins with an imaginative leap that at first seems insane, a leap that begins with 'sudden starts of imagination, which may be termed, perhaps, the ravings of a violent Feaver' (Savinien 1687: 13).

130 Unlike truly fantastic extraordinary voyages such as Savinien's *Histoire des Etats et Empires du Soleil* (1662), translated as the *The Comical History of the States and Empires of the World of the Moon* (1687), in which an adventurer travels between France, New France (Québec) and the Moon in a home-made, dew-powered machine.

131 Other attempts to disguise publication details include false translation notes. For example, Coyer's *A Discovery of the Island Frivolia* (1750) was identified as a translation into English, although it was originally published in English, and appeared in French translation the following year (1751) (Dunmore 1988).

132 Compared to others, Foigny was a little tongue-in-cheek in his truth claims. Vairasse, for example, soberly presented 'very strong Arguments to establish the truth of this History' (Vairasse 1675: A3), and distanced his narrative from other utopias and adventures that were presented as factual, but were generally known to be fictitious. He was quite successful in this; *The History of the Sevarites* was generally received as a true story (Atkinson 1966: 92).

133 A more modern translation of Hall's original Latin is provided by Wands (1981).

134 Although it was the first utopian extraordinary voyage set in Australia, Foigny's narrative was not the first utopia set in Australia, and not the first imaginary voyage to Australia.

135 The island setting of *Swiss Family Robinson* (Wyss 1814), inhabited by kangaroos and 'near to Van Diemen's Land', is superficially Australian (Birmingham and Jeans 1983; Marryat 1841: vii–viii).

136 Although the portion of Australia that could genuinely be portrayed as *terra*

incognita progressively shrank, Australia continued to present relatively large blank spaces on European maps, and continued to accommodate imaginative utopias and dystopias. As the coastline was mapped, it became less plausible for adventurers to be shipwrecked and washed up on the coast of a real but unknown Australia, so Australian imaginary voyages were displaced towards the still-uncharted Australian interior (Gibson 1984).

137 Specifically, Foigny's geographical sources included two cosmographies by Gaston Jean-Baptiste, Baron de Renty, published in Paris in 1645 and 1657 (Fausett 1993).

138 The Australian coast remained a vague outline until the latter part of the eighteenth century when, beginning with Cook's first Australian voyage in 1770, and continuing with a sequence of French and (mostly) English navigators, including George Bass, Matthew Flinders, and Philip Parker King, it was accurately and systematically charted.

139 Willem Jansz was the first European to reach the coast of Australia and document the encounter, in 1605; earlier voyages remained obscure or secret (McIntyre 1977). Jansz left a map but no journal. In 1616, Dick Hartog reached the west coast of Australia. The only evidence of his voyage was a pewter plate he left there. In 1623 Jan Carstensz's second ship reached (what is now) Arnhem Land, but the journal of that voyage was lost and the findings were not incorporated into maps until after 1642. François Pelsaert commanded *The Batavia*, which was wrecked off Australia's west coast in 1629, and suffered a bloody mutiny, described by Pelsaert in a narrative published in 1647. Abel Tasman sighted (what he named) Van Diemen's Land and Staten Land (the South Island of New Zealand) in 1642, when his findings were added to contemporary maps. Written accounts of Tasman's journey were not published until 1671 in Dutch, 1682 in English (Cowley 1988).

140 The first convincing European 'discovery' of (part of what is now) Australia was that of the Spaniard Luis Van de Torres in 1595, on an expedition led by Pedro Fernandes de Queiros, who reached the New Hebrides, which he took for Australia. Queiros published details of his travels in 'memorials' published between 1607 and 1617 (Cowley 1988).

141 Although there were speculations, rumours and reports about the richness of the interior. Spanish explorer Ferdinand de Quiros, for example, stated that Australia was 'more fertile and populous' than any European country (Berneri 1950: 190).

142 Christaller and Weber imagined simplified spaces uniformly populated by 'economic' men, whose locational decisions regarding the service centres they patronise and the places they choose for their manufacturing businesses, respectively, are based on a desire to minimise transport costs (Weber 1929).

143 Said (1993) examines narratives of resistance in European literature, particularly British novels.

7 UNMAPPING ADVENTURES

144 Shakespeare is thought to have been inspired by stories of a ship's crew and passengers, lost and feared drowned in a tempest near Bermuda, who miraculously survived and returned to Europe (Langbaum 1987). Shakespeare completed *The Tempest* in 1611.

145 In *Badlands*, Kroetsch unmaps the Albertan geography inscribed in the field note book and maps of a fictional adventurer, a man who embarked upon a quest into the interior of Alberta's Badlands in search of dinosaur bones. *Badlands* is a dual narrative, the journal of the Dawe Expedition, on the one hand, and the story of

Anna Dawe, retracing her father's journey, on the other. Whereas Dawe's original narrative writes a journey into the interior, mapping its geography, his daughter's narrative unwrites that journey, unmaps its geography.

146 Hunt referred to Orwell in his keynote address to the Conference, 'Imperialism and Gender: Constructions of Masculinity in Twentieth-Century Narrative', organised by the Gender Studies Seminar Group, University of Birmingham, 27 May 1995.

147 The terms post-colonial and post-colonialism are problematic and contested, as McClintock (1994) explains.

148 E.M. Forster (1962) noted in an introduction to *Lord of the Flies* that Golding knew what his pupils were reading.

149 *The Coral Island* was one of the first adventure stories with an all-boy cast (Bratton 1990).

150 Golding also made reference to other literary works, from Homer's *Odyssey* to Swift's *Gulliver's Travels*.

151 Piggy to Ralph.

152 Ballantyne's story is remembered as one of the most representative and influential of the 1850s, in social and cultural if not literary histories (Kermode 1958).

153 Only after British acclaim did the Golding 'vogue' spread to other parts of the world, including the USA (where it was initially allowed to go out of print) and Europe (in translation) (Kearns 1963).

154 Golding's appeal, in the mid-1950s, was to a generation of Britons who had lived through the war, some as soldiers, others at home, and who had seen 'international hatred and . . . totalitarianism' and had perhaps experienced a 'loss of values and of faith' in western civilisation (Grande 1963: 458).

155 Although I do not regard adventures, novels and fables as mutually exclusive categories.

156 Golding's religious comment was much more sophisticated and ambiguous than Ballantyne's. The boys do not pray, and Golding does not moralise. Rather, Golding tells a fable about original sin and the fall of man, in a Garden of Eden. More generally, Hynes (1960: 673) observes, 'religiousness informs the fabric of [Golding's] novels'.

157 Whether they are forced on colonial readers, for example in schools, or whether they are actively sought out and enjoyed by colonial readers, Robinsons and Robinsonades represented colonial encounters to readers all over the world. *Robinson Crusoe* was, as Hazlitt remarked in 1840, a 'powerful influence' throughout 'the nations of Christendom' (Shinagel 1975: 271). Robinsonades such as *The Coral Island* and *Treasure Island* were favourites throughout the English-speaking world, and also beyond it, until after the Second World War. These were the two favourite books among boys surveyed in New Zealand in 1947, for example (Lyons and Taksa 1992). Reading and studying works such as *Robinson Crusoe* and *The Tempest* at the colonial 'peripheries' has the effect of 'reiterating for the colonised the original capture of his/her alterity and the processes of its annihilation and marginalisation' (Brydon and Tiffin 1993: 49–50).

158 Post-colonial literature, neither a post-war nor a recent development, is as old as colonial literature. Ashcroft, Griffiths and Tiffin (1989: 2) 'use the term "post-colonial" . . . to cover all the culture affected by the imperial process from the moment of colonisation to the present day'. Swift was not the first critic to subvert *Robinson Crusoe*, although his was the first critical Robinson to become a best seller. The first critical Robinsons were published months rather than years after the original. Charles Gildon, a minor playwright and pamphleteer, helped invent the critical Robinson by writing *The Life and Strange Surprizing Adventures of Mr. D—— De F—* (1719), a parody of *Robinson Crusoe*. In Gildon's story, D—— De

F— is visited by Robinson Crusoe and Friday, who are seeking revenge for their treatment at the hands of the author. Crusoe is annoyed at being made 'to ramble over three Parts of the World after I was sixty five', while Friday complains of being made to look a 'Blockhead' in his use of the English language, which he picked up after a couple of months, but never (in twelve years of Crusoe's company) succeeded in mastering (Gildon 1719: ix). The two *dramatis personae* threaten D——— De F— with pistols, and force him to eat his own words (literally, to swallow pages of his own work) until they pass right through him, and an 'unsavoury Stench' is emitted from the vicinity of his breeches (Gildon 1719: xviii). Gildon criticised Defoe for his tolerance of slavery, for his populist style, and for the liberties he took with 'facts' (see also Dottin 1923).

159 *Gulliver's Travels* (1726) was much more than an attack on *Robinson Crusoe*, of course, and for much of the eighteenth century it outsold that book (Watt 1957).

160 A starboard tack, for example, cannot be got aboard; it is a point of sail.

161 Although he lived in Ireland for many years.

162 Other literary references in *Moses Ascending* include George Lamming's *Water with Berries* (1971) and Selvon's *The Lonely Londoners* (1956) (Fabre 1979).

163 The children's edition is shorter, simpler and less sexually explicit. I emphasise the children's version of Tournier's novel because, as a juvenile adventure story, it is closest to the heart of my discussion.

164 The British artist Aubrey Beardsley (1872–1898) 'became for many the very symbol of 1890s effeteness and decadence' (Rosenblum and Janson 1984: 458).

165 The dates of Crusoe's birth, shipwreck and rescue, according to Defoe and Tournier respectively, were 30 Sept. 1632 and 19 Dec. 1737; 30 Sept. 1659 and 30 Sept. 1759; 19 Dec. 1686 and 19 Dec. 1787 (Purdy 1984).

166 Coetzee refers to *Robinson Crusoe* and explores and negotiates race relations and colonialism in previous works, notably the novel *Dusklands*, but *Foe* is his first explicit and sustained response to *Robinson Crusoe*. Gardiner (1987: 174) calls *Dusklands* a South African 'translation of *Robinson Crusoe*' and argues that Coetzee's works 'have all subversively inscribed Daniel Defoe's *Robinson Crusoe* with the deliberate aim of rejecting its canonical formulation of the colonial encounter'.

167 Susan Barton shares her name with the heroine of Helen Dore Boylston's series that included such titles as *Sue Barton, Student Nurse* (1989), which were read mainly in Britain and the USA in the 1930s and 1940s.

168 Barton's adventures, including her rape, are reminiscent of those of the rare eighteenth-century female Crusoes such as Hannah Hewit (Dibdin 1790).

169 As one critic put it, Barton's 'essence is that she is a teller of tales', who 'carries the central theme of Coetzee's *Foe*, the nature of narrative art' (Penner 1989: 116).

170 Donoghue (1987: 26) remarks that Coetzee's narrator 'has evidently been reading Jacques Derrida's *De La Grammatologie*'.

CONCLUSION

171 The Vintage edition of Said's *Culture and Imperialism* (1993) is labelled an 'international bestseller' (front cover). Hyam's work on empire, particularly on sexual life in the British Empire, found its way from readable academic texts to watchable prime-time television documentary. Hyam was the consultant for the BBC2 television series *Ruling Passions*, first screened in February, March and April 1995, accompanied by the book *Ruling Passions: Sex, Race and Empire* (Gill 1995).

172 The modern parents refer to books originally published two or three generations ago but still popular, by Johns (1924), Burroughs (1917) and Hergé (1931).

173 Said (1993) takes a similar stance with respect to more 'serious' cultural forms that have also been associated with imperialism, but are not necessarily intrinsically imperialist.

174 There is, of course, still a children's entertainment market, in which books, film and television products, and now video and computer games, are targeted specifically at children.

175 Adventure books continue to be produced at both popular and 'serious' ends of the market and in between (Bristow 1995). Exploration adventures, for example, remain a recurring motif in Canadian and Australian literature (Atwood 1972; Huggan 1994).

176 Despite the apparent media saturation of the late twentieth century, the President of the Association of American Geographers was able to claim, in 1985, that the general public still have an unquenched thirst for geographical information. He suggested that geographers learn to master new technologies and create vivid images, in film and video, in maps and on computer displays, and in 'vivid attractive English' (Lewis 1985: 465).

177 Felix Driver drew my attention to advertisements by River Island, a clothing retailer and brand name, in which the RGS logo appears alongside images of rugged-looking men in adventurous outdoor settings, helping to sell the menswear product as an accessory of the rugged, masculine man.

178 The image of Europeans transplanting – simplifying and purifying – their culture overseas is echoed in Canadian stories and histories ranging from Jack Hodgins' *The Invention of the World* (1977), the story of an Irishman who transplants an entire village to Vancouver Island where he builds the Revelations Colony of Truth, to Cole Harris's more formal thesis about the simplification of Europe overseas in which some elements of Europe were brought to Canada while others were left behind (Harris 1977; see also Hartz 1964).

179 Brydon and Tiffin (1993), borrowing Macherey's (1978) concept of Robinson Crusoe as 'thematic ancestor', call Patrick White's *Voss* a 'thematic' descendant of Joseph Conrad's *Heart of Darkness* and in turn of *Robinson Crusoe.*

180 Selvon uses Robinson Crusoe as a point of departure from which to get from a colonial to a post-colonial condition; the language and geography of a colonial story is his point of departure for writing post-colonial (West Indian) literature (King 1979; Morris 1984).

181 The rational 'utopia' of German planner and geographer Walter Christaller, for example, was conceived in a German military hospital, where the wounded soldier gazed at maps and imagined a less complicated world emerging from the rubble of war. Back on the front line, Christaller kept the pocket atlas with him, holding onto the images of a simpler world, images which helped him fight, to play a part in the violent struggle for a new world in Europe (Christaller 1972).

BIBLIOGRAPHY

Abramson, H.S. (1987) *National Geographic: Behind America's Lens on the World*, New York: Crown Publishers.

Aitchison, R.C. (1909) 'Jules Verne, 1828–1905', *The Boy's Own Paper* 31: 557–558.

Allen, W.O.B. and McClure, E. (1898) *Two Hundred Years, the History of the SPCK, 1698–1898*, Chicago.

Alpers, S. (1983) *The Art of Describing*, Chicago: University of Chicago Press.

Andre (1881) *Robinson Crusoe; or, De Friend, De Foe, and De Foetprint, a pantomimic medley*, London: Theatre Royal, Drury Lane.

Andrews, S.K. (1989) review of *Cartography in the Media*, in *The American Cartographer* 16.

Anon. (attributed to John Holmesby) (1757) *The Voyages, Travels, and Wonderful Discoveries of Captain John Holmesby, Containing a Series of the most Surprising and Uncommon Events which Befel the Author in his Voyage to the Southern Ocean, in the Year 1739*, London: Noble.

Anon. (1857) *The Welsh Family Crusoes; or, The Lonely Island*, London: Simpkin, Marshall.

Anon. (undated a) *The Wonderful Life and Surprising Adventures of Robinson Crusoe*, London: William Darton.

Anon. (undated b) *One Farthing Robinson Crusoe*, London: R. March.

Arnold, G. (1980) *Held Fast for England: G.A. Henty, Imperialist Boys' Writer*, London: Hamish Hamilton.

Artibise, A.F.J. (1978) *Western Canada Since 1870*, Vancouver: University of British Columbia Press.

Ashcroft, B., Griffiths, G. and Tiffin, H. (1989) *The Empire Writes Back: Theory and Practice in Post-colonial Literatures*, London: Routledge.

—— (1995) *The Post-colonial Studies Reader*, London: Routledge.

Atkinson, R.G. (1966) *The Extraordinary Voyage in French Literature Before 1700*, New York: AMS Press / Colombia University.

Atwood, M. (1972) *Survival*, Toronto: Anansi.

Auchmuty, R. (1992) *A World of Girls*, London: Women's Press.

Auerbach, N. (1987) 'A novel of her own', *New Republic* 3764.

Baker, J.N.L. (1931) 'The geography of Daniel Defoe', *Scottish Geographical Magazine* 47: 257–269.

Bakhtin, M.M. (1981) 'Forms of time and of the chronotope in the novel', in M. Holquist (ed.) *The Dialogic Imagination*, Austin: University of Texas Press.

Ballantyne, J. (1834) *The Prose Works of Sir Walter Scott*, Edinburgh.

Ballantyne, R.M. (1848) *Hudson's Bay; or, Every-Day Life in the Wilds of North America*, Edinburgh: William Blackwood.

—— (1856) *Snowflakes and Sunbeams; or, The Young Fur Traders, a Tale of the Far North*, London: T. Nelson.

—— (1858a) *The Coral Island, a Tale of the Pacific Ocean*, London: T. Nelson.

—— (1858b) *Ungava: a Tale of Esquimaux-land*, London: T. Nelson.

—— (1861a) *The Dog Crusoe, a Tale of the Western Prairies*, London: T. Nelson.

—— (1861b) *The Golden Dream; or, Adventures in the Far West*, London: John F. Shaw.

—— (1861c) *The Gorilla Hunters, a Tale of the Wilds of Africa*, London: T. Nelson

—— (1863a) *The Wild Man of the West; a Tale of the Rocky Mountains*, London: Routledge, Warne & Routledge.

—— (1863b) 'Away in the Wilderness; or, Life Among the Red Indians and Fur Traders of North America', *Ballantyne's Miscellany* II, London: James Nisbet.

—— (1869) 'Over the Rocky Mountains; or, Wandering Will in the Land of the Redskins', *Ballantyne's Miscellany* III, London: James Nisbet.

—— (1872) *The Pioneers: a Tale of the Western Wilderness*, London: James Nisbet.

—— (1874) *Tales of Adventure by Flood, Field and Mountain*, London: James Nisbet.

—— (1880) *The Red Man's Revenge; a Tale of the Red River Flood*, London: James Nisbet.

—— (1886) *The Prairie Chief*, London: James Nisbet.

—— (1887) *The Big Otter; a Tale of the Great Nor'West*, London: George Routledge.

—— (1891) *The Buffalo Runners; a Tale of the Red River Plains*, London: James Nisbet.

—— (1893) *Personal Reminiscences in Book Making*, London: James Nisbet.

—— (1954) *The Coral Island*, London: J.M. Dent.

Bann, S. (1990) 'From Captain Cook to Neil Armstrong: colonial exploration and the structure of a landscape', in S. Pugh (ed.) *Reading Landscape: Country–City–Capital*, Manchester: Manchester University Press.

Barnes, J. (1986) 'Through clear Australian eyes', in P.R. Eaden and F.H. Mares (eds) *Mapped But Not Known: The Australian Landscape of the Imagination*, Adelaide: Wakefield Press.

Barr, J. (1986) *Illustrated Children's Books*, London: British Library.

Begg, A. (1871) *Dot it Down. Life in the Northwest*, Toronto: Harper, Rose & Co.

Ben-Amos, I.K. (1995) *Adolescence and Youth in Early Modern England*, New Haven: Yale University Press.

Berneri, M.L. (1950) *Journey Through Utopia*, London: Freedom Press.

Bhabba, H. (1983) 'The other question', *Screen* 24, 6: 18–36.

Billington, M. (1993) 'Cannibals and converts', *Manchester Guardian Weekly*, 17 January.

Birkett, D. (1991) *Spinsters Abroad: Victorian Lady Explorers*, London: Victor Gollancz.

Birmingham, J.M. and Jeans, D.N. (1983) '*The Swiss Family Robinson* and the archaeology of colonisations', *Australian Historical Archaeology* 1.

Blaeu, J. (1663) *Le Grand Atlas*, Amsterdam.

Blair, D. (1882) 'The first imaginary voyage to Australia', *Victorian Review* 7: 199–204.

Blunt, A. (1994) *Travel, Gender and Imperialism: Mary Kingsley and West Africa*, New York: Guilford Press.

Blunt, A. and Rose, G. (eds) (1994) *Writing Women and Space: Colonial and Postcolonial Geographies*, New York: Guilford Press.

Bolt, C. (1995) 'Guides for the Pure', *The Times Higher Education Supplement*, 17 February: 25.

Bonyhady, T. (1991) *From Melbourne to Myth*, Balmain, New South Wales: David Ell Press.

Borrow, G. (1851) *Lavengro: the Scholar – the Gypsy – the Priest* (3 volumes), London: J. Murray.

Bott, T.H. (1882) *Robinson Crusoe in Verse*, London: Simpkin Marshall.

Boyd, K. (1991) 'Knowing your place: the tensions of manliness in boys' story papers,

1918-39', in M. Roper and J. Tosh (eds) *Manful Assertions: Masculinities in Britain since 1800*, London: Routledge.

Boylston, H.D. (1949) *Sue Barton, Student Nurse*, London: John Lane.

Boy's Champion Story Paper, The (1903) 'Correspondence', 21 February.

Bradbury, R. (1990) 'Foreword', in W. Butcher, *Verne's Journey to the Centre of the Self: Space and Time in the Voyages Extraordinaires*, London: Macmillan.

Brantlinger, P. (1988) *Rule of Darkness: British Literature and Imperialism, 1830–1914*, Ithaca, New York: Cornell University Press.

Bratton J.S. (1981) *The Impact of Victorian Children's Fiction*, London: Croom Helm.

—— (1989) 'British imperialism and the reproduction of femininity in girls' fiction, 1900–1930', in J. Richards (ed.) *Imperialism and Juvenile Literature*, Manchester: Manchester University Press.

—— (1990) 'Introduction', in R.M. Ballantyne, *The Coral Island*, Oxford: Oxford University Press.

—— (1991) *Acts of Supremacy: the British Empire and the Stage*, Manchester: Manchester University Press.

Bristow, J. (1991) *Empire Boys: Adventures in a Man's World*, London: HarperCollins.

—— (ed.) (1995) *The Oxford Book of Adventure*, Oxford: Oxford University Press.

Brown, J.S.H. (1980) *Strangers in Blood: Fur Trade Families in Indian Country*, Vancouver: UBC Press.

Brown, L.A. (1977) *The Story of Maps*, New York: Dover.

Brydon, D. and Tiffin, H. (1993) *Decolonising Fictions*, Sydney: Dangaroo Press.

Burn, I. (1980) 'Beating about the bush: the landscapes of the Heidelberg School', in A. Bradley and T. Smith (eds) *Australian Art and Architecture: Essays Presented to Bernard Smith*, Melbourne: Oxford University Press.

Burroughs, E.R. (1917) *Tarzan of the Apes*, New York.

Butcher, W. (1990) *Verne's Journey to the Centre of the Self: Space and Time in the Voyages Extraordinaires*, London: Macmillan.

Cadogan, M. and Craig, P. (1986) *You're a Brick Angela! The Girls' Story 1839–1985*, London: Victor Gollancz.

Camm, J.C.R. and McQuilton, J. (eds) (1988) *Australians: a Historical Atlas, 1888*, Sydney: Fairfax, Syme & Weldon.

Campbell, J. (1949) *The Hero with a Thousand Faces*, New York: Bollingen/Pantheon.

Carter, P. (1987) *The Road to Botany Bay: an Exploration of Landscape and History*, New York: Alfred Knopf.

Chesneaux, J. (1972) *The Political and Social Ideas of Jules Verne*, trans. T. Wikeley, London: Thames & Hudson.

Christaller, W. (1972) 'How I discovered the theory of central places', in P.W. English and R. Mayfield (eds) *Man, Space and Environment*, New York: Oxford University Press.

Clarke, M. (1876) Preface to A.L. Gordon, *Sea Spray and Smoke Drift*, Melbourne: Clarson, Massina.

Clune, F. (1937) *Dig! a Drama of Central Australia*, Sydney: Angus & Robertson.

Coetzee, J.M. (1986) *Foe*, London: Secker & Warburg.

Compère, D. (1974) 'M. Jules Verne conseiller municipal', in P. Tourrain (ed.) *Jules Verne*, Paris: Editions de l'Herne, 127–140.

Comus (R.M. Ballantyne) (1856) *Three Little Kittens*, Edinburgh: T. Nelson.

—— (1857) *Mister Fox*, Edinburgh: T. Nelson.

Conrad, J. (1899) 'Heart of Darkness', *Blackwood's Edinburgh Magazine* CLXV, February–April.

—— (1926) 'Geography and some explorers', *Last Essays*, London: J.M. Dent. Reprint of 1924 pamphlet of the same title, London: Strangeways.

Cooper J.F. (1827) *The Prairie, A Tale* (in three volumes), Paris: Hector Bossange.

Costello, P. (1978) *Jules Verne: Inventor of Science Fiction*, London: Hodder & Stoughton.

Cowley, D. (1988) 'European voyages of discovery', *La Trobe Library Journal* 11, 41: 15–20.

Cox, C.B. (1960) 'Lord of the Flies', *Critical Quarterly* 2, 2: 112–117.

Cox, J. (1982) *Take a Cold Tub, Sir! The Story of the Boy's Own Paper*, Guildford: Lutterworth Press.

Coyer, A. (1750) *A Discovery of the Island Frivolia; or, The Frivolous Island*, London: T. Payne.

Crush, J. (1994) 'Post-colonialism, de-colonization, and geography', in A. Godlewska and N. Smith (eds) *Geography and Empire*, Oxford: Blackwell.

Dahl, E. (1977) *Die Kurzungen des 'Robinson Crusoe' in England Zwischen 1719 und 1819*, Frankfurt: Peter Lang (Anglo-American Forum Series).

Darwin, C. (1859) *On the Origin of Species by Means of Natural Selection, or the Preservation of Favoured Races in the Struggle for Life*, London: John Murray.

Daunt, A. (1882) *The Three Trappers; a Story of Adventure in the Wilds of Canada*, London: Nelson.

—— (1885) *In the Land of Moose, Bear and Beaver; Adventures in the Forests of Athabasca*, London: Nelson.

Davey, F. (1979) 'The explorer in western Canadian literature', *Studies in Canadian Literature* 4, 2.

Dawson, G. (1994) *Soldier Heroes: British Adventure, Empire and the Imagining of Masculinities*, London: Routledge.

Day, H. (1967) 'Autour de Louise Michel et de Jules Verne', *Défense de l'Homme* 219: 29–32.

Defoe, D. (1719a) *The Life and Strange and Surprizing Adventures of Robinson Crusoe, of York, Mariner: who lived eight and twenty years, all alone in an uninhabited island on the coast of America, near the mouth of the great river of Oroonoque; having been cast on shore by shipwreck, whereon all the men perished but himself. With an account of how he was at last as strangely deliver'd by pyrates. Written by himself*, London: W. Taylor.

—— (1719b) *The Farther Adventures of Robinson Crusoe; being the second and last part of his life, and of the strange surprizing accounts of his travels round three parts of the globe. Written by himself. To which is added a map of the world, in which is delineated the voyages of Robinson Crusoe*, London: W. Taylor.

—— (1720) *Serious Reflections during the Life and Surprising Adventures of Robinson Crusoe; with his vision of the angelick world. Written by himself*, London: W. Taylor.

—— (1725) *A New Voyage Round the World by a Course Never Sailed Before*, London: A. Bettesworth.

—— (1886) *Robinson Crusoe: His Life and Adventures, after Daniel Defoe, illustrated with 48 chromolithographs*, London: SPCK.

Deleuze, G and Guattari, F. (1988) *A Thousand Plateaus: Capitalism and Schizophrenia*, London: Athlone.

Demos, J. and Demos, J. (1969) 'Adolescence in historical perspective', *Journal of Marriage and the Family* 31.

Dibdin, C. (1790) *Hanna Hewit; or, the Female Crusoe. Being the History of a Woman of Uncommon Mental and Personal Accomplishments; who, After a Variety of Extraordinary and Interesting Adventures in Almost Every Station of Life, from Splendid Prosperity to Abject Adversity, was Cast Away in the Grosvenor East-Indianman; and Became for Three Years the Sole Inhabitant of an Island in the South Seas. Supposed to be written by herself*, London: C. Dibdin.

Dixon, R.J. (1986) *The Course of Empire: Neoclassical Culture in New South Wales, 1788–1860*, Melbourne: Oxford University Press.

—— (1995) *Writing the Colonial Adventure: Race, Gender and Nation in Anglo-Australian Popular Fiction*, Cambridge: Cambridge University Press.

Domosh, M. (1991) 'Towards a feminist historiography of geography', *Transactions of the Institute of British Geographers New Series* 16: 95–104.

Donaldson, M. (1993) 'What is hegemonic masculinity?', *Theory and Society* 22: 643–657.

Donoghue, D. (1987) 'Her man Friday', *New York Times Book Review*, 22 February.

Dottin, P. (1923) *Robinson Crusoe Examin'd and Criticis'd; or, a New Edition of Charles Gildon's Famous Pamphlet now Published with an Introduction and Explanatory Notes*, London: J.M. Dent.

Doyle, B. (1968) *The Who's Who of Children's Literature*, London: Hugh Evelyn.

Driver, F. (1992) 'Geography's empire: histories of geographical knowledge', *Environment and Planning D, Society and Space* 10: 23–40.

—— (1994) 'Geography triumphant? Joseph Conrad and the imperial adventure', *The Conradian* 18, 2: 103–111.

Dunae, P.A. (1976) '*Boy's Own Paper*: origins and editorial policies', *The Private Library* 9, 2, 4: 123–158.

—— (1989) 'New Grub Street for boys', in J. Richards (ed.) *Imperialism and Juvenile Literature*, Manchester: Manchester University Press.

Dunbar, G.S. (1973) 'Isotherms and politics: perception of the northwest in the 1850s', in R. Allen (ed.) *Prairie Perspectives 2*, Toronto: Holt, Rinehart & Winston.

Duncan, J. and Duncan, N. (1988) '(Re)reading the landscape', *Environment and Planning D, Society and Space* 6: 117–126.

Dunmore, J. (1988) *Utopias and Imaginary Voyages to Australasia*, Canberra: National Library of Australia.

Dyhouse, C. (1981) *Girls Growing up in Late Victorian and Edwardian England*, London: Routledge & Kegan Paul.

Eagleton, T. (1990) *The Ideology of the Aesthetic*, Oxford: Basil Blackwell.

Eaton, S. (1969) *Lady of the Backwoods: a Biography of Catharine Parr Traill*, Toronto: McClelland & Stewart.

Egoff, S.A. (1992) *Canadian Children's Books 1799–1939, a Bibliographical Catalogue*, Vancouver: University of British Columbia Library.

—— and J. Saltman (1990) *The New Republic of Childhood*, Toronto: Oxford University Press.

Evans, I.O. (ed.) (1964a) *The Children of Captain Grant Part I: The Mysterious Document*, London: Arco.

—— (ed.) (1964b) *The Children of Captain Grant Part II: Among the Cannibals*, London: Arco.

—— (1965) *Jules Verne and his Work*, London: Arco.

Fabre, M. (1979) 'Samuel Selvon', in B. King (ed.) *West Indian Literature*, London: Macmillan.

Faivre, J.P. (1969) 'Les voyages extraordinaires de Jules Verne en Australie', *Australian Journal of French Studies* 5: 205–221.

—— (1970) 'L'Australie et les Enfants du Capitaine Grant', *Bulletin de la Société Jules Verne* 13: 100–104.

Fardell, J. (1995) 'The modern parents', *Viz* 71, April/May: 21–22.

Fausett, D. (1993) *Writing the New World: Imaginary Voyages and Utopias of the Great Southern Land*, Syracuse: Syracuse University Press.

Favenc, E. (1881) *The Great Austral Plain, its Past, Present and Future*, Sydney: H.R. Woods.

—— (1888) *The History of Australian Exploration from 1788 to 1888*, Sydney: Turner & Henderson.

—— (1896a) *The Secret of the Australian Desert*, London: Blackie.

—— (1896b) *Marooned on Australia; the Narrative by Diedrich Buys of his Discoveries and Exploits in Terra Australis Incognita About the Year 1630*, London: Blackie.

—— (1896c) *The Moccasins of Silence*, Sydney: George Robertson.

—— (1898) *The New Standard Geography of Australasia*, Sydney: George Robertson.

—— (1902) *The Geographical Development of Australia*, Sydney and Brisbane: William Brooks.

—— (1908) *The Explorers of Australia, and their Life-work*, London and Melbourne: Whitcombe & Tombs.

Fleming, T. (1969) 'Friday', *The New York Times Book Review*, 13 April.

Flood, G.D. (ed.) (1993) *Mapping Invisible Worlds*, Edinburgh: University of Edinburgh Press.

Foigny, G. (1676) *La Terre Australe connue: C'est à Dire, La Description de ce pays inconnu jusqu'ici, de ses moeurs & de ses coûtumes. Par M. Sadeur. Avec les avantures qui le conduisirent en ce Continent, & les particularitez du sejour qu'il y fit durant trente-cinq ans & plus, & de son retour*, Vannes (Geneva): Jacques Verneuil.

—— (1693) *A New Discovery of Terra Incognita Australis, or the Southern World, by James Sadeur, a French-Man who Being Cast there by a Shipwrack, lived 35 years in that Country, and gives a particular Description of the Manners, Customs, Religion, Laws, Studies, and Wars, of those Southern People; and of some Animals peculiar to that Place*, London: John Dunton.

—— (1993) *The Southern Land, Known*, trans. D. Fausett, Syracuse: Syracuse University Press.

Forster, E.M. (1962) 'Introduction', in W. Golding, *Lord of the Flies*, New York: Coward-McCann.

Foucault, M. (1978) *The History of Sexuality Volume I, An Introduction*, Harmondsworth: Penguin.

Fox, Lady M. (attributed to the Reverend R. Whately) (1837) *Account of an Expedition to the Interior of New Holland; The Late Wonderful Discovery of a Civilized Nation of European Origin, Which Had, in so Remarkable a Manner, Been Kept Separate Hitherto from the Rest of the Civilized World*, London: Richard Bentley.

Francis, R.D. (1989) *Images of the West*, Saskatoon: Western Prairie Producer Press.

Francke, L. (1995) 'Danger – Boy Zone: boys and their toys, muscles and tensions, and barely a woman in sight. Lizzie Francke lets herself be swept away by sleek, all-action movies', *The Guardian*, Saturday 28 October: 26.

Friederich, W.P. (1967) *Australia in Western Imaginative Prose Writings, 1600–1960*, Chapel Hill: University of North Carolina Press.

Friesen, G. (1984) *The Canadian Prairies: a History*, Toronto: University of Toronto Press.

Frost, C. (1983) 'The last explorer: the life and work of Ernest Favenc', *Foundations for Australian Literary Studies Monograph 9*, Townsville, Australia: James Cook University of North Queensland.

Frye, N. (1990) *Anatomy of Criticism*, Harmondsworth: Penguin Books.

Gallagher, E.J., Mistichelli, J.A. and Van Eerde, J.A. (1980) *Jules Verne: a Primary and Secondary Bibliography*, Boston: G.K. Hall.

Garagnon, J. (1981) 'French imaginary voyages to the Austral lands in the seventeenth and eighteenth centuries', in I. Donaldson (ed.) *Australia and the European Imagination*, Canberra: Australian National University.

Gardiner, A. (1987) 'J.M. Coetzee's *Dusklands*: colonial encounters of the Robinsonian kind', *World Literature Written in English* 27, 2: 174–184.

Gerrard, B. (1988) 'Introduction: the Great South Land', *La Trobe Library Journal*, 11, 41: 3–7.

Gibson, R. (1984) *The Diminishing Paradise: Changing Literary Perceptions of Australia*, London: Angus & Robertson.

—— (1992) *South of the West: Postcolonialism and the Narrative Construction of Australia*, Bloomington: Indiana University Press.

Gildon, C. (1719) *The Life and Strange Surprizing Adventures of Mr. D—De F-, of London, Hosier. In a dialogue between him, Robinson Crusoe, and his man Friday*, London: J. Roberts

Giles, F. (1988) 'Romance: an embarrassing subject', in L. Hergenhan (ed.) *The Penguin New Literary History of Australia*, Melbourne: Penguin, 223–237.

Gill, A. (1995) *Ruling Passions: Sex, Race and Empire*, London: BBC Books.

Gillies, P. (1994) *Shakespeare and the Geography of Difference*, Cambridge: Cambridge University Press.

Godolphin, M. (1868) *Robinson Crusoe in Words of One Syllable*, London: George Routledge.

Golding, W. (1954) *Lord of the Flies*, London: Faber & Faber.

Gove, P.B. (1975) *The Imaginary Voyage in Prose Fiction*, New York: Columbia University Press.

Grande, L. (1963) 'The appeal of Golding', *Commonweal* LXXVII, 25 January.

Gray, M.A. (1992) *A Dictionary of Literary Terms*, London: Longman.

Green, M. (1979) *Dreams of Adventure, Deeds of Empire*, New York: Basic Books.

—— (1990) *The Robinson Crusoe Story*, Pennsylvania: Pennsylvania State University Press.

—— (1991) *Seven Types of Adventure Tale: an Etiology of a Major Genre*, Pennsylvania: Pennsylvania State University Press.

Gregory, D. (1994a) 'Geographical imagination', in R.J. Johnston, D. Gregory and D.M. Smith (eds) *The Dictionary of Human Geography*, Oxford: Blackwell.

—— (1994b) *Geographical Imaginations*, Oxford: Blackwell.

Haggard, H.R. (1885) *King Solomon's Mines*, London: Cassell.

Hakluyt, R. (1598–1600) *The Principal Navigations, Voyages and Discoveries of the English Nation, Made by Sea or Over Land*, London: Bishop & Newberie.

Haley, B. (1978) *The Healthy Body and Victorian Culture*, Cambridge: Harvard University Press.

Hall, G.S. (1904) *Adolescence: its Psychology and its Relationship to Anthropology, Sociology, Sex, Crime, Religion and Education*, New York.

Hall, J. (1605) *Mundus Alter et Idem, Sive Terra Australis Ante Hac Semper Incognita Longis Itineribus Peregrini Academici Nuperrime Lustrata, Auth: Mercurius Britannicus*, Frankfurt.

—— (1609) *The Discovery of a New World, or, a Description of the South Indies, Hitherto Unknowne, by an English Mercury*, London.

—— (1981) *Another World and Yet the Same; or, the Southern Continent, Before this always Unknown, through the Extended Travels of a Wandering Academic most Recently Surveyed*, trans. J.M. Wands, New Haven: Yale University Press.

Hammerton, A.J. (1979) *Emigrant Gentlewomen*, London: Croom Helm.

Hannabuss, S. (1983) 'Islands as metaphors', *Universities Quarterly* 38, 1: 70–82.

Hardyment, C. (1984) *Arthur Ransome and Captain Flint's Trunk*, London: Jonathan Cape.

Harley, J.B. (1992) 'Deconstructing the map', in T. Barnes and J. Duncan (eds) (1992) *Writing Worlds: Discourse, Text and Metaphor in the Representation of Landscape*, London: Routledge.

Harper, J.R. (1971) *Paul Kane 1810–1871*, Ottawa: National Gallery.

Harris, R.C. (1977) 'The simplification of Europe overseas', *Annals of the Association of American Geographers*, 67, 4.

—— (1982) 'Regionalism and the Canadian archipelago', in L.D. McCann (ed.) *Heartland and Hinterland. A Geography of Canada*, Scarborough, Ontario: Prentice-Hall.

Harrison, D. (1977) *Unnamed Country: the Struggle for a Canadian Prairie Fiction*, Edmonton: University of Alberta Press.

—— (1980) 'Popular fiction of the Canadian prairies: autopsy on a small corpus', *Journal of Popular Culture* 14, 2.

Hartz, L. (1964) *The Founding of New Societies*, New York: Harcourt, Brace & World.

Hergé (1931) *Tintin en Amérique*, Paris: Casterman; L. Lonsdale-Cooper and M. Turner (trans.) (1978) *Tintin in America*, London: Methuen.

Hodgins, J. (1977) *The Invention of the World*, Toronto: Macmillan.

Hodgson, M. (1967) 'The exploration journal as literature', *The Beaver* 298: 4–12.

Homer (1995) *The Odyssey*, trans. A.T. Murray, revised G.E. Dimock, London: Harvard University Press.

Hooper, M. (1982) *Doctor Hunger and Captain Thirst: Stories of Australian Explorers*, Sydney: Methuen Australia.

Horner, J.C. (1911) 'A dark tragedy – a thrilling story of western life', unpublished journal, Saskatchewan Archives Board Regina R–E2278, Regina, Canada.

Huggan, G. (1989a) 'Decolonizing the map: post-colonialism, post-structuralism and the cartographic connection', *Ariel* 20, 4: 115–131.

—— (1989b) 'Maps and mapping strategies in contemporary Canadian and Australian fiction', unpublished dissertation, University of British Columbia.

—— (1994) *Territorial Disputes*, Toronto: University of Toronto Press.

Hughes, T. (1857) *Tom Brown's School Days; By an Old Boy*, Cambridge: Macmillan.

Hulme, P. (1992) *Colonial Encounters: Europe and the Native Caribbean 1492–1797*, London: Routledge.

Hyam, R. (1990) *Empire and Sexuality*, Manchester: Manchester University Press.

—— (1993) *Britain's Imperial Century, 1815–1914*, Basingstoke: Macmillan.

Hynes, S. (1960) 'Novels of a religious man', *Commonweal* LXXI, 18 March: 673–675.

Jackel, S. (ed.) (1982) *A Flannel Shirt and Liberty*, Vancouver: UBC Press.

Jackson, D. (1990) *Unmasking Masculinity: a Critical Autobiography*, London: Unwin Hyman.

Jackson, P. (1989) *Maps of Meaning*, London: Unwin Hyman.

—— (1991) 'The cultural politics of masculinity: towards a social geography', *Transactions, Institute of British Geographers, New Series* 16: 199–213.

—— (1994) 'Black male: advertising and the cultural politics of masculinity', *Gender, Place and Culture* 1, 1.

—— (1995) 'Postcolonialism', in R.J. Johnston, D. Gregory and D.M. Smith (eds) *The Dictionary of Human Geography*, Oxford: Blackwell.

Jameson, F. (1988) 'Cognitive mapping', in C. Nelson and L. Grossberg (eds) *Marxism in the Interpretation of Culture*, Illinois: University of Illinois Press.

Jenkinson, A.J. (1946) *What do Boys and Girls Read?*, London: Methuen.

Johns, W.E. (1942) *Biggles Flies East*, London: Hodder & Stoughton.

Johnson, J.W. (1981) 'The utopian impulse and southern lands', in I. Donaldson (ed.) *Australia and the European Imagination*, Canberra: Australian National University.

Joyce, J. (1964) 'Daniel Defoe', trans. J. Prescott, *Buffalo Studies* I, 1.

Kane, P. (1859) *Wanderings of an Artist Among the Indians of North America from Canada to Vancouver's Island and Oregon Through the Hudson's Bay Company Territory and Back Again*, London: Longman, Brown, Green, Longmans & Roberts.

Katz, C. and Smith, N. (1993) 'Grounding metaphor: towards a spatialized politics', in M. Keith and S. Pile (eds) *Place and the Politics of Identity*, London: Routledge, 67–83.

Katz, W. (1987) *Rider Haggard and the Fiction of Empire*, Cambridge: Cambridge University Press.

Kearns, F.E. (1963) 'Salinger and Golding: conflict on the campus', *America* CVIII, 26 January: 136–139.

Keltie, J.S. (1907) 'Fictitious travel and phantom lands', *Harper's Monthly Magazine* 54, July: 186–194.

Kermode, F. (1958) 'Coral Islands', *The Spectator* CCI, 22 August: 257.

—— (1961) 'The novels of William Golding', *International Literary Annual* III: 11–29.

Kestner, J.A. (1995) *Masculinities in Victorian Painting*, Vermont and Aldershot: Scolar Press.

Kett, J. (1971) 'Adolescence and youth in nineteenth-century America', *Journal of Interdisciplinary History* 2.

—— (1977) *Rites of Passage: Adolescence in America, 1790 to the Present*, New York: Basic Books.

King, B. (ed.) (1979) *West Indian Literature*, London: Macmillan.

Kingsford, Rev. M.R. (1947) *The Life, Work and Influence of William Henry Giles Kingston*, Toronto: Ryerson.

Kingston, W.H.G. (1871) *Captain Cook, His Life, Voyages and Discoveries*. London: RTS.

—— (1879) *Frontier Fort; or, Stirring Times in the North-West Territory of British North America*, London: SPCK.

—— (1881) *Adventures in the Far West*, London: G. Routledge.

—— (1882) *Arctic Adventures*, London: G. Routledge.

—— (1883) *Adventures in Africa, by an African Trader*, London: G. Routledge.

—— (1884a) *Australian Adventures*, London: G. Routledge.

—— (1884b) *Adventures in India*, London: G. Routledge.

Knox-Shaw, P. (1987) *The Explorer in English Fiction*, Basingstoke: Macmillan.

Koeman, C. (1970) *Joan Blaeu and his Grand Atlas*, London: George Philip.

Kolodny, A. (1975) *The Lay of the Land*, Chapel Hill: University of North Carolina Press.

Kroetsch, R. (1975) *Badlands*, Toronto: New Press.

—— (1989) *The Lovely Treachery of Words*, Toronto: Oxford University Press.

Kuipers, B. (1982) 'The map in the head metaphor', *Environment and Behaviour* 14: 202–220.

Lachèvre, F. (1922) *Les Successeurs de Cyrano de Bergerac*, Paris: Librairie Ancienne Honoré Champion.

Lamming, G. (1971) *Water with Berries*, Port of Spain: Longman Caribbean; London: Longman.

Landow, G. (1982) *Images of Crisis: Literary Iconology, 1750 to the Present*, London: Routledge & Kegan Paul.

Lang, A. (1891) *Essays in Little*, London: Whitefriars Library of Wit and Humour.

Langbaum, R. (ed.) (1987) *The Tempest*, New York: Signet.

Lansbury, C. (1970) *Arcady in Australia: the Evocation of Australia in Nineteenth-century English Literature*, Melbourne: Melbourne University Press.

Lean, G. (1995) 'Explorers hail 1990s as "golden age of discovery"' *Independent on Sunday*, 30 April: 14.

Lees, C. (ed.) (1994) *Medieval Masculinities: Regarding Men in the Middle Ages*, Minneapolis and London: University of Minnesota Press.

Leighly, J. (ed.) (1963) *Land and Life: a Selection from the Writings of Carl Ortwin Sauer*, Berkeley: University of California Press.

Lewis, P. (1985) 'Presidential address: beyond description', *Annals of the Association of American Geographers* 75, 4: 465–477.

Livingstone, D.N. (1992) *The Geographical Tradition: Episodes in the History of a Contested Enterprise*, Oxford: Blackwell.

Longueville, P. (1727) *The Hermit; or, the Unparalleled Sufferings and Surprising Adventures of Mr. Philip Quarll, an Englishman, who was Lately Discovered by Mr. Dorrington, a Bristol Merchant, upon an Uninhabited Island in the South Sea; Where he has Lived about Fifty Years, Without any Human Assistance, Still Continues to Reside and Will not Come Away*, Westminster: T. Warner.

Lowther Clarke, W.K. (1959) *A History of the SPCK*, London: SPCK.

Lyons, M. and Taksa, L. (1992) *Australian Readers Remember*, Melbourne: Oxford University Press.

McClintock, A. (1994) *Imperial Leather: Race, Gender and Sexuality in the Colonial Contest*, New York and London: Routledge.

Macdonald, A.C. (1907) *The Lost Explorers: a Story of the Trackless Desert*, London: Blackie.

—— (1915) 'Central Australia – its possibilities', *Victoria Geographical Journal* XXXI, II: 4–19.

McDougall (1895) *Forest, Lake and Prairie; 20 Years of Frontier Life in Western Canada*, Toronto: W. Briggs.

Macgregor, F. (1989) *Robert Louis Stevenson*, Norwich: Jarrold.

Macherey, P. (1978) *A Theory of Literary Production*, trans. G. Wall, London: Routledge.

McIntyre, K. (1977) *The Secret Discovery of Australia*, Menindie, South Australia: Souvenir Press.

Mackannes, G. (1937) 'Some fictitious voyages to Australia', *Royal Australian Historical Society: Journal and Proceedings* 23: 153–159.

Mackenzie, J.M. (ed.) (1986) *Imperialism and Popular Culture*, Manchester: Manchester University Press.

Mackie, J. (1901) *The Heart of the Prairie*, London: George Newnes.

—— (1912) *A Bush Mystery; or, the Lost Explorer*, London: Nisbet.

McLanathan, R. (1968) *The American Tradition in the Arts*, London: Studio Vista.

MacLulich, T.D. (1977) 'The explorer as hero: Mackenzie and Fraser', *Canadian Literature* 75: 61–73.

—— (1979) 'Canadian exploration as literature', *Canadian Literature* 81: 72–85.

Major, A. (1991) 'Bessie Marchant, the maid of Kent whose exciting stories thrilled thousands of English children', *This England*, Winter: 30–34.

Mangan, J.A. (1986) *The Games Ethic and Imperialism: Aspects of the Diffusion of an Ideal*, Harmondsworth: Viking.

Mann, M. (1986) *The Sources of Social Power, Volume I, A History of Power from the Beginning to AD 1760*, Cambridge: Cambridge University Press.

Marchant, B. (1906a) *A Daughter of the Ranges: a Story of Western Canada*, London: Blackie.

—— (1906b) *Athabasca Bill: a Tale of the Far West*, London: Christian Knowledge Society.

—— (1907) *Sisters of Silver Creek: a Story of Western Canada*, London: Blackie.

—— (1908) *Courageous Girl: a Story of Uruguay*, London: Blackie.

—— (1909) *Daughters of the Dominion: a Story of the Canadian Frontier*, London: Blackie.

—— (1911) *A Countess from Canada. A story of Life in the Backwoods*, London: Blackie.

—— (1912) *The Ferry House Girls: an Australian Story*, London: Blackie.

—— (1914) *A Mysterious Inheritance: a Story of Adventure in British Columbia*, London: Blackie.

—— (1917) *A Canadian Farm Mystery; or, Pam the Pioneer*, London: Blackie.

—— (1918) *Cynthia Wins; a Tale of the Rocky Mountains*, London: Blackie.
—— (1919) *Norah to the Rescue: a Story of the Philippines*, London: Blackie.
—— (1920) *Sally Makes Good: a Story of Tasmania*, London: Blackie.
—— (1921) *The Girl of the Pampas*, London: Blackie.
—— (1923) *Rachel Out West*, London: Blackie.
Marryat, C.F. (1834) *Mr. Midshipman Easy* (3 volumes), London: Saunders & Otley.
—— (1841) *Masterman Ready; or, The Wreck of the Pacific*, London: Longman.
Martin, A. (1985) *The Knowledge of Ignorance from Genesis to Jules Verne*, Cambridge: Cambridge University Press.
Marx, L. (1964) *The Machine in the Garden*, New York: Oxford University Press.
Masterman, N.C. (1963) *John Malcolm Ludlow, the Builder of Christian Socialism*, Cambridge: Cambridge University Press.
Mein, S.G. (1985) 'The Aberdeen Association: an early attempt to provide library services to settlers in Saskatchewan', *Saskatchewan History* 38, 1: 2–19.
Meinig, D.W. (1962) *On the Margins of the Good Earth*, Chicago: Rand McNally/ Association of American Geographers.
Melville, H. (1846) *Typee; or, a Narrative of a four months' residence among the natives of a valley of the Marquesas islands; or, a peep at Polynesian life*, London: John Murray.
—— (1851) *The Whale*, London; simultaneously published in the United States of America as *Moby Dick*.
Milham, M.E. (1964) 'Introduction', in J.F. Cooper, *The Prairie*, New York: Airmont.
Milton, Viscount W.W.F. and Cheadle, W.B. (1865) *The North-West Passage by Land: being the narrative of an expedition from the Atlantic to the Pacific undertaken with the view of exploring a route across the continent to B.C. through British territory, by one of the northern passes of the Rocky Mountains*, London: Cassell, Petter & Galpin.
Montgomery, L.M. (1908) *Anne of Green Gables*, London: Sir Isaac Pitman.
Montgomery, S.L. (1993) 'Through a lens brightly: the world according to *National Geographic*', *Science as Culture* 4.
Moodie, S. (1852) *Roughing it in the Bush; or, Life in Canada*, London.
—— (1853) 'Lost Children', in *Life in the Clearings Versus the Bush*, London.
Moore, J.R. (1941) 'The geography of *Gulliver's Travels*', *Journal of English and Germanic Philology* 40: 214–228.
—— (1958) *Daniel Defoe: Citizen of the Modern World*, Chicago.
Moorehead, A. (1963) *Cooper's Creek*, New York: Atlantic Monthly Press.
Moors, D. (1988) 'Imaginary voyages', *La Trobe Library Journal* 11, 41: 10.
Moreland, C. and Bannister, D. (1986) *Antique Maps*, Oxford: Phaidon/Christies.
Morris, M. (1984) 'Introduction', in S. Selvon, *Moses Ascending*, London: Heinemann.
Moyles, R.G. and Owram, D. (1988) *Imperial Dreams and Colonial Realities: British Views of Canada 1880–1914*, Toronto: University of Toronto Press.
Mullan, B. (1987) *The Enid Blyton Story*, London: Boxtree/TVS.
Nelson, C. (1991) *Boys Will Be Girls: the Feminine Ethic and British Children's Fiction 1857–1917*, New Brunswick and London: Rutgers University Press.
Nerlich, M. (1987) *Ideology of Adventure, Studies in Modern Consciousness, 1150–1750* (in two volumes), Minneapolis: University of Minnesota Press.
Neville, H. (1668) *The Isle of Pines; or, a Late Discovery of a Fourth Island Near Terra Australis Incognita, by Henry Cornelius Van Sloetten, Wherein is Contained a True Relation of Certain English Persons, who in Queen Elizabeth's Time, Making a Voyage to the East Indies Were Cast Away, and Wrecked Near the Coast of Terra Australis Incognita, and all Drowned, Except One Man and Four Women, and how Lately Anno. Dom. 1667 a Dutch Ship . . . by Chance have Found their Posterity, (Speaking Good English,) to Amount, (as They Suppose,) to Ten or Twelve Thousand Persons*, London: Allen Banks & Charles Harper.

New, W.H. (1972) *Articulating West*, Toronto: New Press.

Newbery, J. (1765) *The History of Little Goody Two-Shoes; otherwise called Mrs. Margery Two-Shoes*, London: St Paul's Church Yard.

Niall, B. (1988) 'Children's Literature', in L. Hergenhan (ed.) *The Penguin New Literary History of Australia*, Melbourne: Penguin.

Niemeyer, C. (1961) '*The Coral Island* revisited', *College English* XXII, January: 241–245.

Nordenskiold, A.J. (1973) *Facsimile-Atlas to the Early History of Cartography*, New York: Dover.

Oldsley, B. and Weintraub, S. (1965) *The Art of William Golding*, New York: Harcourt, Brace & World.

Orwell, G. (1940) 'Boys' weeklies', in *Inside the Whale and Other Essays*, London: Victor Gollancz.

Owram, D. (1980) *Promise of Eden*, Toronto: University of Toronto Press.

Oxley, J.M. (1892) *Fergus MacTavish, or, A Boy's Will, a Story of the Far North West*, Philadelphia: American Baptist Publishing Co.

—— (1893) *Archie of Athabasca*, Boston: Lothrop. Reprinted in 1908 as *The Young Nor'Wester*, London: Religious Tract Society.

—— (1896) *The Boy Tramps; or, Across Canada*, Toronto: Musson; London, W. & R. Chambers.

—— (1901) *North Overland with Franklin*, London: Religious Tract Society.

Packer, G. (1987) 'Blind alleys', *Nation* 224, 28 March.

Palliser, J. (1859) *Papers Relative to the Exploration by Captain Palliser of that Portion of British North America Which Lies Between the Northern Branch of the River Saskatchewan and the Frontier of the United States, and Between the Red River and Rocky Mountains, Presented to Both Houses of Parliament by Command of Her Majesty*, London: Eyre & Spottiswood.

Palmer, V. (1954) *The Legend of the Nineties*, Melbourne: Melbourne University Press.

Parker, C. (1991) 'Race and empire in the stories of R.M. Ballantyne', in R. Giddings (ed.) *Literature and Imperialism*, New York: St Martin's Press.

Pearlman, E. (1976) '*Robinson Crusoe* and the cannibals', *Mosaic* 10, 1: 39–55.

Peel, B. (1968) 'English writers of the early west', *Alberta Historical Review*, 16, 2: 1–4.

Penner, D. (1989) *Countries of the Mind: the Fiction of J.M. Coetzee*, New York: Greenwood Press.

Phillips, R.S. (1993) 'The language of images in geography', *Progress in Human Geography* 17, 2: 180–194.

Pile, S. and Thrift, N. (eds) (1995) *Mapping the Subject: Geographies of Cultural Transformation*, London: Routledge.

Post, J.B. (1979) *An Atlas of Fantasy*, New York: Ballantine.

Powell, J.M. (1988) *An Historical Geography of Modern Australia*, Cambridge: Cambridge University Press.

Pratt, M.L. (1992) *Imperial Eyes: Travel Writing and Transculturation*, London: Routledge.

Purdy, A. (1984) 'From Defoe's *Crusoe* to Tournier's *Vendredi*: the metamorphosis of a myth', *Canadian Review of Comparative Literature* 9: 216–223.

Quayle, E. (1967) *Ballantyne the Brave: a Victorian Writer and his Family*, London: Rupert Hart-Davis.

—— (1968) *R.M. Ballantyne: A Bibliography of First Editions*, London: Rupert Hart-Davis.

Raban, J. (1969) 'Inventing worlds', *New Society* 13, 332: 217–218.

Ransome, A. (1930) *Swallows and Amazons*, London: Jonathan Cape.

Ray, S.G. (1982) *The Blyton Phenomenon, the Controversy Surrounding the World's Most Successful Children's Writer*, London: André Deutsch.

Rees, R. (1988) *New and Naked Land: Making the Prairies Home*, Saskatoon: Western Prairie Producer Press.

Reid, M. (1853) *The Boy Hunters; or, Adventures in Search of a White Buffalo*, London: David Bogue.

Reynolds, K. (1990) *Girls Only? Gender and Popular Children's Fiction in Britain, 1880–1910*, Philadelphia: Temple University Press.

—— and Humble, N. (1995) *Victorian Heroines: Representations of Femininity in 19th-century Literature and Art*, London: Harvester Wheatsheaf.

Richards, J. (ed.) (1989) *Imperialism and Juvenile Literature*, Manchester: Manchester University Press.

Riffenburgh, B. (1993) *The Myth of the Explorer*, London: Wiley.

Rigney, J. (1991) 'Preface', *The Isle of Pines*, Katoomba, N.S.W.: Wayzgoose Press.

Rogers, P. (1979) *Robinson Crusoe*, London: George Allen & Unwin.

Roper, E. (1891) *By Track and Trail, A Journey Through Canada*, London: W.H. Allen.

Roper, M. and Tosh, J. (eds) (1991) *Manful Assertions: Masculinities in Britain since 1800*, London: Routledge.

Rose, G. (1993) *Feminism and Geography: the Limits of Geographical Knowledge*, Minneapolis: University of Minnesota Press.

Rosenblum, R. and Janson, H.W. (eds) (1984) *Art of the Nineteenth Century*, London: Prentice-Hall.

Rothenberg, T. (1994) 'Voyeurs of imperialism: *The National Geographic Magazine* before World War II', in A. Godlewska and N. Smith (eds) *Geography and Empire*, Oxford: Blackwell.

Rousseau, J.J. (1762) *Emilius and Sophia; or, a New System of Education*, London.

Ruskin, J. (1865) *Sesame and Lilies: Two Lectures, Delivered at Manchester in 1864*, London: Smith, Elder.

Said, E. (1993) *Culture and Imperialism*, London, Vintage.

Salmon, E. (1886a) 'What the working classes read', *The Nineteenth Century*, 20.

—— (1886b) 'What girls read', *The Nineteenth Century*, 20: 515–529.

—— (1888) *Juvenile Literature As It Is*, London: Henry Drane.

—— and Longden, A.A. (1924) *The Literature of the Empire*, London: W. Collins.

Sarup, M. (1994) 'Home and Identity', in G. Robertson, M. Mash, L. Tickner, J. Bird, B. Curtis and T. Putnam (eds) *Travellers' Tales: Narratives of Home and Displacement*, London: Routledge, 93–104.

Sauer, C.O. (1925) 'The morphology of landscape', *University of California Publications in Geography*, 2, 2: 19–53, reprinted in J. Leighly (ed.) (1963) *Land and Life: a Selection from the Writings of Carl Ortwin Sauer*, Berkeley: University of California Press.

—— (1966) *The Early Spanish Main*, Berkeley: University of California Press.

Savage, A. (1994) 'The story's voyage through the text: transformations of the narrative in *Beowulf*' in K. Pratt (ed.) *Shifts and Transpositions in Medieval Narrative*, Cambridge: D.S. Brewer.

—— (1995) 'Anglo-Saxons, Anglo-Saxonists, men and women', paper presented, Medieval Masculinities Conference, University of Cardiff, 25 November.

Savinien, C. (1662) *Histoire des Etats et Empires du Soleil*, Paris.

—— (1687) *The Comical History of the States and Empires of the World of the Moon*, trans. A. Lovell, London: Henry Rhodes.

Saxby, C.F.A. (1907a) *The Call of Honour; a Tale of Adventure in the Canadian Prairies*, London: S.W. Partridge.

—— (1907b) *Braves, White and Red; a Tale of Adventures in the North-west*. London and Edinburgh: T.C. & E.C. Jack.

—— (1921) *The Settler of Serpent Creek, A Tale of the Canadian Prairie*, London: Boy's Own Paper Office.

Saxby, J. (1889) *Kate and Jean; the History of Two Young and Independent Spinsters*, Edinburgh: Oliphant, Anderson & Ferrier.
—— (1890) *West-Nor'West*, London: J. Nisbet.
—— (1896) *Brown Jack: a Tale of North West Canada*, London: Religious Tract Society.
Schaffer, K.F. (1988) *Women and the Bush: Forces of Desire in the Australian Cultural Tradition*, Sydney: Cambridge University Press.
Schieder, R. (ed.) (1986) *Canadian Crusoes - a Tale of the Rice Lake Plains*, Ottawa: Carleton University Press/Centre for Editing Early Canadian Texts.
Schofield, R.S. (1981) 'Dimensions of illiteracy in England 1750-1850', in H.J. Graff (ed.) *Literacy and Social Development in the West*, Cambridge: Cambridge University Press.
Sedgwick, E.K. (1985) *Between Men: English Literature and Male Homosocial Desire*, New York: Columbia University Press.
Seidel, M. (1991) *Robinson Crusoe: Island Myths and the Novel*, Boston: Twayne.
Selby, J. (1963) 'Ballantyne and the fur traders', *Canadian Literature* 18: 40–46.
Selvon, S.D. (1956) *The Lonely Londoners*, London: Allan Wingate.
—— (1975) *Moses Ascending*, London: Davis-Poynter.
Serle, G. (1973) *From Deserts the Prophets Come*, Melbourne: Heinemann.
Shields, R. (1991) *Places on the Margin: Alternative Geographies of Modernity*, London: Routledge.
Shinagel, M. (1975) *Robinson Crusoe*, New York: Norton.
Shohat, E. (1991) 'Imagining terra incognita: the disciplinary gaze of empire', *Public Culture* 3, 2: 41–70.
Short, J.R. (1991) *Imagined Country: Environment, Culture and Society*, London: Routledge.
Silver, A.I. (1969) 'French Canada and the prairie frontier, 1870–1890', *Canadian History Review* 1, 1.
Skelton, R.A. (1952) *Decorative Printed Maps of the 15th to 18th Centuries*, London: Spring Books.
—— (1968) 'The cartography of Columbus' first voyage', in C. Columbus, *The Journal of Christopher Columbus*, trans. C. Jane, London: Anthony Blond.
—— (1975) *Maps: a Historical Survey of the Study and Collecting*, Chicago and London: University of Chicago Press.
Smith, N. (1994) 'Geography, empire and social theory', *Progress in Human Geography* 18, 4: 491–500.
Soja, E. (1989) *Postmodern Geographies: the Reassertion of Space in Social Theory*, London: Verso.
Somekawa, E. and Smith, E.A. (1988) 'Theorising the writing of history', *Journal of Social History*, 22.
Spark, M. (1958) *Robinson*, New York: Avon.
Springhall, J.O. (1973) 'Rule Britannia! Hope and glory for armchair imperialists', *The British Empire* 55: 1513–1540, London/Nederland: BBC/Time-Life.
Stanley, J. (ed.) (1995) *Bold in her Breeches: Women Pirates Across the Ages*, London: Pandora.
Stead, R.J.C. (1914) *The Bail Jumper*, London: T. Unwin Fisher.
—— (1927) *Grain*, New York: G.H. Doran.
Stephen, L. (1874) 'De Foe's Novels', in *Hours in A Library*, London.
Stevenson, K. (1958) 'Jules Verne in Victoria', *Southerly* 19, 1: 23–25.
Stevenson, R.L. (1883) *Treasure Island*, London: Cassell.
—— (1962) *Treasure Island*, New York: Airmont.
Stock, B. (1986) 'Texts, readers and enacted narratives', *Visible Language* 20: 294–301.
Stone, L. (1969) 'Literacy and education in England, 1640–1900', *Past and Present* 42.
Strachey, E.J. (1959) *The End of Empire*, London: Gollancz.

Street, B.V. (1975) *The Savage in Literature: Representations of 'Primitive' Society in English Fiction, 1858–1920*, London: Routledge & Kegan Paul.

Strickland, A. (ed.) (1852) Foreword in C.P. Traill (1852) *Canadian Crusoes: a Tale of the Rice Lake Plains*, London: Arthur Hall, Virtue.

Suvin, D. (1974) 'Communication in quantified space: the utopian liberalism of Jules Verne's science fiction', *Clio* 4: 51–71.

Swift, J. (1726) *Travels into Several Remote Nations of the World, in Four Parts, by Lemuel Gulliver, First a Surgeon and then a Captain of Several Ships*, London: Benj. Motte.

Taussat, R. (1974) 'L'anarchisme divin: de l'Isle Lincoln à l'Isle Hoste', in P.A. Touttain (ed.) *Jules Verne*, Paris: Editions de l'Herne, 242–255.

Thiong'o, N. (1986) *Decolonizing the Mind*, London: James Currey.

Thompson, D. (ed.) (1995) *The Concise Oxford Dictionary*, 9th edn, Oxford: Clarendon Press.

Thomson, P. (1956) *The Victorian Heroine, a Changing Ideal 1837–73*, London.

Times Literary Supplement (1972) 'The Bourgeois Facade', 3689 (17 November): 1391.

Tinkler, P. (1995) *Constructing Girlhood: Popular Magazines for Girls Growing up in England 1920–1950*, London: Taylor & Francis.

Torgovnick, M. (1990) *Gone Primitive: Savage Intellects, Modern Lives*, Chicago: University of Chicago Press.

Tournier, M. (1969) *Friday; or The Other Island*, trans. N. Denny, New York: Pantheon.

—— (1972) *Friday and Robinson; Life on Speranza Island*, New York: Knopf.

Traill, C.P. (1846) *The Backwoods of Canada*, London: M.A. Nattali.

—— (1852) *Canadian Crusoes: a Tale of the Rice Lake Plains*, London: Arthur Hall, Virtue.

—— (1854) *The Female Emigrant's Guide, and Hints on Canadian Housekeeping* (second edition), Toronto.

—— (1895) *Canadian Wild Flowers. Painted and lithographed by Agnes Fits Gibbon, with botanical descriptions by C.P. Traill* (first published in 1868), Toronto: W. Briggs.

—— (1986) *Canadian Crusoes: a Tale of the Rice Lake Plains*, Ottawa: Carleton University Press/Centre for Editing Early Canadian Texts.

Tucker, N. (1995) 'Martyrs to Mickey', *The Times Higher Education Supplement*, 23 June: 33.

Tulloch, C. (1959) 'Pioneer reading', *Saskatchewan History* 12, 3: 97–99.

Turner, E.S. (1957) *Boys Will Be Boys*, London: Michael Joseph.

Turner, V. (1969) *The Ritual Process*, Chicago: Aldine.

—— (1982) *From Ritual to Theatre: the Human Seriousness of Play*, New York: Performing Arts Journal Publications.

Ullrich, H. (1898) *Robinson und Robinsonaden*, Weimar.

Vairasse, D. (1675) *The History of the Sevarites, or Sevarambi: a Nation Inhabiting Part of the Third Continent, Commonly Called, Terrae Australes Incognitae, with an Account of their Admirable Government, Religion, Customs, and Language. Written by one Captain Siden, a Worthy Person, who, Together with many Others, was Cast upon those Coasts, and Lived many Years in that Country*, London: Henry Brome.

Vance, N. (1985) *The Sinews of the Spirit*, Cambridge: Cambridge University Press.

Van den Abbeele, G. (1980) 'Sightseers: the tourist as theorist', *Diacritics* 10: 2–14.

Van Gennep, A. (1909) *Les Rites de Passage*, Paris.

—— (1960) *The Rites of Passage*, trans. M.B. Vizedom and G.L. Caffee, London: Routledge & Kegan Paul.

Verne, J. (1863) *Cinq Semaines en ballon: Voyages de découvertes en Afrique*, Paris: J. Hetzel.

—— (1864) *Voyage au centre de la terre*, Paris: J. Hetzel.

—— (1865) *De la Terre à la Lune: Trajet direct en 97 heures*, Paris: J. Hetzel.

—— (1866a) *Les Anglais au Pole Nord: Voyages et aventures du Capitaine Hatteras* (Part 1), Paris: J. Hetzel.

—— (1866b) *Le Desert de glace: Voyages et aventures du Capitaine Hatteras* (Part 2), Paris: J. Hetzel.

—— (1867–8) *Les Enfants du Capitaine Grant* (three volumes), Paris: J. Hetzel.

—— (1869–70) *Vingt Mille Lieues sous les mers* (two volumes), Paris: J. Hetzel.

—— (1877) *A Voyage Round the World* (three volumes), London and New York: George Routledge.

—— (1964a) *The Children of Captain Grant Part 1: The Mysterious Document*, London: Arco.

—— (1964b) *The Children of Captain Grant Part 2: Among the Cannibals*, London: Arco.

Vincent, D. (1989) *Literacy and Popular Culture in England 1750–1914*, Cambridge: Cambridge University Press.

Wackermann, E. (1976) 'Robinson und Robinsonaden: buch illustration aus drei jahrhunderten', *Illustration* 63.

Walvin, J. (1982) *A Child's World: a Social History of English Childhood 1880–1914*, Harmondsworth: Penguin.

Wands, J. M. (trans., ed.) (1981) *Another World and Yet the Same; or, the Southern Continent, Before this always Unknown, through the Extended Travels of a Wandering Academic most Recently Surveyed*, New Haven: Yale University Press.

Ware, V. (1992) *Beyond the Pale: White Women and History*, London: Verso.

Warren, S. (1993) 'This heaven gives me migraines: the problems and promise of landscapes of leisure', in David Ley and James Duncan (eds) *Place/Culture/Representation*, London: Routledge.

Watt, I. (1951) '*Robinson Crusoe* as a myth', *Essays in Criticism* 1, 2: 95–119.

—— (1957) *The Rise of the Novel*, Berkeley and Los Angeles: University of California Press.

Weber, A. (1929) *Theory of the Location of Industries*, Chicago: University of Chicago Press.

Whalley, J.I. and Chester, T.R. (1988) *A History of Children's Book Illustration*, London: John Murray.

Wheelwright, J. (1987) '"Amazons and military maids": an examination of female military heroines in British literature and the changing construction of gender', *Women's Studies International Forum* 10, 5: 489–502.

—— (1989) *Amazons and Military Maids: Women who Dressed as Men in the Pursuit of Life, Liberty and Happiness*, London: Pandora.

White, A. (1993) *Joseph Conrad and the Adventure Tradition: Constructing and Deconstructing the Imperial Subject*, Cambridge: Cambridge University Press.

White, E. (1993) *Genet*, London: Picador.

White, H. (1973) *Metahistory: the Historical Imagination in Nineteenth-Century Europe*, Baltimore: Johns Hopkins University Press.

White, P.V.M. (1957) *Voss*, London: Eyre & Spottiswoode.

Woods, G. (1995) 'Fantasy islands: popular topographies of marooned masculinity', in D. Bell and G. Valentine (eds) *Mapping Desire*, London: Routledge, 126–148.

Woolf, V. (1932) 'Robinson Crusoe', in *The Common Reader, Second Series*, London: Harcourt Brace Jovanovich.

Wright, J.K. (1947) 'Presidential address: Terrae incognitae: the place of the imagination in geography', *Annals of the Association of American Geographers* 37, 1: 1–15.

Wyss, J.D. (1814) *The Swiss Family Robinson; or, Adventures of a Father and a Mother and Four Sons on a Desert Island*, London: William Goodwin.

BIBLIOGRAPHY

Yonge, C.M. (1886) *What Books to Lend and What to Give,* London: National Society's Depository.
Young, E.R. (1899) *Winter Adventures of Three Boys in the Great Lone Land,* London: Robert Cullen.
Zweig, P. (1974) *The Adventurer,* London: Dent.

INDEX

Note: page nos in *italics* refer to plates; those in **bold** refer to the major reference for an item.

202

T - #0176 - 071024 - C0 - 234/156/12 - PB - 9780415137720 - Gloss Lamination